MORPHOLOGY OF FUNGI

By

Shubhrata R. Mishra

DPH

DISCOVERY PUBLISHING HOUSE
NEW DELHI

First Published-2005

ISBN 81-7141-980-1

Published by

DISCOVERY PUBLISHING HOUSE
4831/24, Ansari Road, Prahlad Street,
Darya Ganj, New Delhi-110002 (India)
Phone: 23279245 • Fax: 91-11-23253475
E-mail:dphtemp@indiatimes.com

Printed at:

Amit Enterprises, Delhi

Preface

Fungi are thallophytes that have no greed plant pigment. Their body is generally called thallus, is either a single cell or a thread like structure. We can distinguish several hundred fungus species by the naked eye, but this has its limitations.

This book is written to throw light on the world of microcosm lavishly abounding in extra ordinary and aesthetically impressive forms opens before us.

I hope the students of B.Sc and M.Sc. Agriculture, would get more benefit from this book.

I wish to express my sincere thanks to the publisher, Mr. Tilak Wasan for his continued interest in bringing out this edition.

I shall regard my efforts amply rewarded if the book is received with the enthusiasm.

Author

Contents

1

Introduction

Fungi are thallophytes that have no green plant pigment. Fungi are generally reproduced by spores. The Fungi derive organic substances from the bodies of other organisms, living or dead.

The fungal body which is generally called thallus is either a single cell or a thread like structure. That is called hyphae.

In most of fungi, spore producing cells, form a part of special structure made up of hyphal tissue and called the fruit body (sporocarp).

Although the fruit bodies are given to, extraordinary variation in shape, size and colouring these do remain fairly constant in individual fungus groups. The same applies to spores. It is exactly the internal and external morphology of fruit bodies.

The simplest types of fungi possess a unicellular, filamentous, branching thallus, whose cytoplasm contains, a great member of partitions to divide them.

Under specific conditions, the fungi produce sporiferous cells.

In many fungi the fruit bodies are the first to develop and subsequently to originate spore bearing cells.

2

The Fruit Body

The fruit body is an extraordinary multiform structure in its dimensions (ranging from microscopic size up to the size of several tens of centimetres) and also in its shape and colouring.

The fruit body of simpler forms may be nothing but a little ball of hyphae entwining the asci.

In ascomycetes the fruit body is either globular permanently closed (in which case it is known as cleistothecium) or it opens either through a regular pore or irregularly.

Finally the apothecium the fruit body characteristic of the discomycetes is a more or less open cup shaped or discoid fruit body.

Now let us study about the two types of gymnocarpous fruit bodies.

1. Resupinate Fruit Bodies

These fruit bodies form more or less conspicuous layers with either indistinct or sharply delimited margins of various consistency, usually very thin but in some cases several millimetres thick. Basidia as a rule are arranged in to a hymenium covering the entire upper surface of the fruit body. On the substrate it is often turned downwards to the ground. Resupinate fruit bodies may be found on the underside of dropped branches and uprooted trunks. Sometimes a tendency to produce lateral caps or horms with a partially elevated margins can be seen.

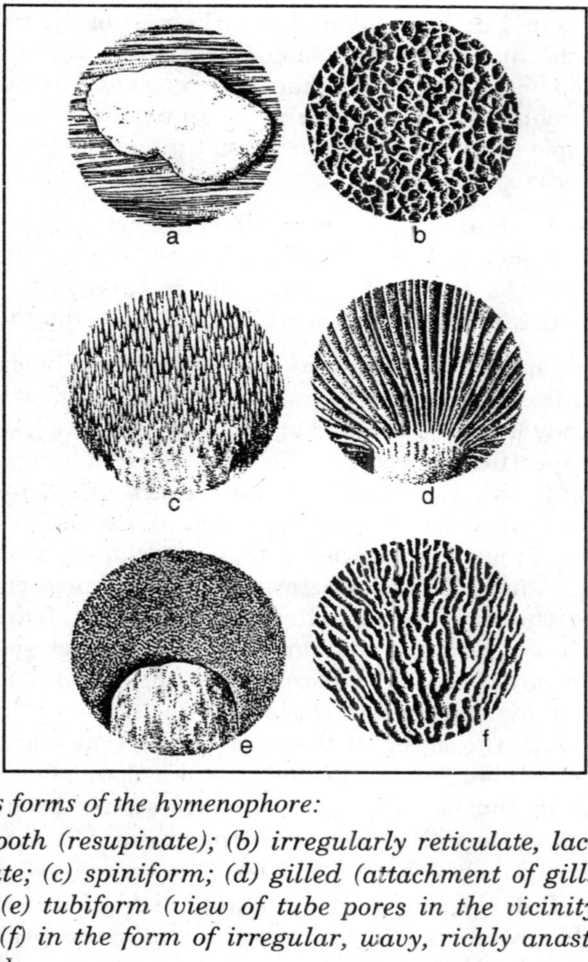

Various forms of the hymenophore:
(a) smooth (resupinate); (b) irregularly reticulate, lacunar to faveolate; (c) spiniform; (d) gilled (attachment of gills to the stipe); (e) tubiform (view of tube pores in the vicinity of the stipe); (f) in the form of irregular, wavy, richly anastomosed gill folds.

2. Upright Fruit Bodies

They occur both in the Basidiomycetes and in some Discomycetes. The hymenium covers either the entire external surface or only its upper part (for example in the Clavariaceae). The most developed type is a fruit body differentiated into a stem or stipe (stipes) and a cap (pileus). Most gill fungi and boleti, as well as some tooth-fungi belong to this group. The hymenophore, i.e. the part of the fruit body bearing the hymenium, is usually confined to the underside of the cap. Stipeless fruit bodies, attached to the substrate either dorsally or laterally, occur in some genera of gill fungi (for example *Crepidotus*) and in a great many polypore species (Polyporaceae, Poriaceae).

The hymenophore takes different forms: it is either completely smooth or variously corrugated (veined, wrinkled), or assumes the shape of spines, gills or tubes. All of these forms occur in all the types of fruit bodies referred to above.

The cap and the stipe are also multiform. The cap often changes its shape in the course of the gradual development of fruit body from youth to old age, and is, moreover, subjected to intraspecific variability, i.e. the caps of single fungi belonging to the same species are not exactly the same. Gills (lamellae) play an important part in classifying and identifying fungi. The manner of their attachment to the stipe is of importance. This characteristic appears most clearly in the longitudinal section through the fresh fruit body, preferably a young and undamaged one—in older specimens it is not always possible to discover how the gills are attached. Of no less significance is the colouring of the gills which changes with the aging of the fruit body. As it matures the fruit body's colouring is affected by the colour of the spores ripening on the basidia which are sometimes produced in quantities so large that they cover the entire gill surface as a white or variously coloured powder. The same applies to the tubes of Boletales and to hymenophores of other fungi. Although nowadays the colouring of the spore print is no longer considered a principal classification criterium of higher fungi, it continues to be of major importance in determining the species.

The cap cuticle provides a great number of important and constant characteristics. Its properties: separability, sliminess, viscosity, dryness, fibrousness, scaliness, etc. are determined by its microscopic structure. More detailed information is given in the chapter dealing with fungi under the microscope. The most conspicuous features is the colouring of the cuticle—it is the first to attract attention. However, the specification of colours is a highly subjective matter. A certain degree of objectivity may be reached by comparing the colours of the fruit body with printed colour tables (known as chromotaxes) where the colours are arranged according to a certain system and numbered, and eventually named. Several such chromotaxes have been worked out—most of them, however, are inaccessible to non-specialists. For example, one of the most accomplished ones, repeatedly cited in all fundamental mycological works, is Ridgeway R., *Colour Standards and Colour Nomenclature,* Published in 1912. In other chromotaxes exactly those hues are often lacking which most frequently appear on fungus fruit bodies, i.e. brown, yellow and red. In most cases, the colour table attached to Moser's key *Die Röhrlinge und Blätterpilze,* 1978, may adequately be used, though to a limited extend.

In gill fungi and boleti, quite young fruit bodies are often enclosed in a cover called a velum. This cover is of two types: the universal veil (*velum universale* or *velum generale*), and the partial veil (cobwebby veil or cortina—*velum partiale*). The former deposits scabby remains on the surface of the cap and a volva on the stipe base of a fully developed fruit body, while the latter connects the cap margin with the stipe and its remnants may be found both in the marginal zone of the cap and in the basal portion of the stipe; also the ring encircling the upper part of the stipe belongs to them. The heterogeneous character of the velum, ranging from fine filaments to a hard membranous cover, makes it essential to examine both quite young and fully developed fruit bodies in identifying some genera (for example *Cortinarius*).

The flesh is of various consistency – it may be very friable, pomaceously brittle, fleshy, watery, succulent, dry, fibrillose, coriaceous, tough or woody. When cut or broken, a number of significant features may be seen of which the most relevant is the colour change which may sometimes be different in the cap and in the stipe, while in the stipe differences may in turn prevail in the colouring of the basal portion and of the other parts. Sometimes the context exudes

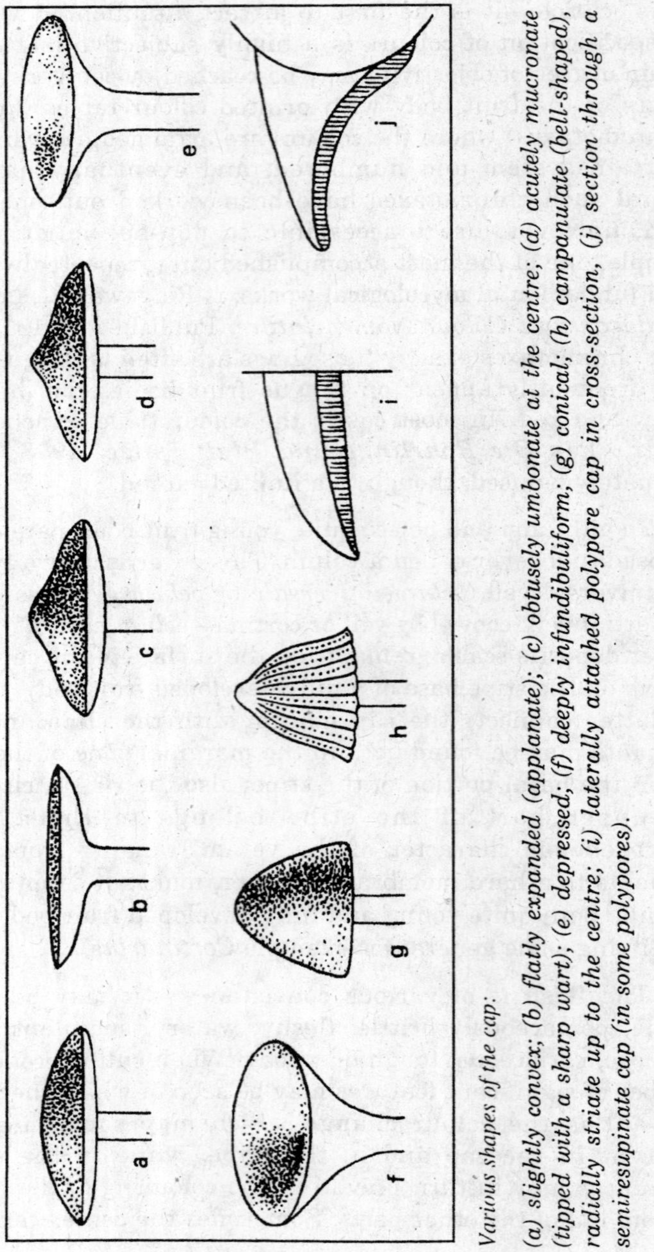

Various shapes of the cap:

(a) slightly convex; (b) flatly expanded (applanate); (c) obtusely umbonate at the centre; (d) acutely mucronate (tipped with a sharp wart); (e) depressed; (f) deeply infundibuliform; (g) conical; (h) campanulate (bell-shaped), radially striate up to the centre; (i) laterally attached polypore cap in cross-section, (j) section through a semiresupinate cap (in some polypores).

latex, a colourless or milky liquid whose colouring is either persistent or changes on exposure to air. No less important is the taste of the context (tasting belongs to the indispensable diagnostic procedures in identifying fungi). A discussion of smells and odours emitted by fruit bodies could fill a chapter by itself. There are many of them, and they are often compared to the well-known, as well as to the less usual, smells characteristic of other organisms, plants, animals, or chemical compounds. Their evaluation is often largely subjective, yet the smells are so specific in the individual species that a good expert is in a position to identify a fungus even from small fragments of the fruit body on the basis of their characteristic smell only. Some fungi (e.g. the tooth fungi) maintain their specific smell even as exsiccations (dried specimens).

3

The Study of Fungi

We can distinguish several hundred fungus species by the naked eye, but this has its limitations, and in a great many cases, the identification remains questionable without confirmation by microscopic analysis.

Consequently the confirmed entusiast should consider whether to invest time and money in the use of a microscope.

If so the fascinating world of microcosm, lavishly abounding in extraordinary and aesthetically impressive forms opens before us.

The actual fungus body (mycelium) is a relatively uniform structure composed of long cylindrical septate colourless or coloured hyphae sometimes provided with clamp connections. The hyphal walls are smooth, and may be thin or, thick and sometimes encrusted.

The differentiation of individual organs of the fungus body becomes plane only in fruit bodies. The most important organ is the hymenophore, whose internal structure remains constant in the individual genera.

In is best observed on the tangential section across the gills, obtained manually with the help of a razor blade.

With a little feeling and patience we are sure to succeed in selecting from among a larger number of available sections at least one sufficiently thin to enable us to observe the structure of the trama in a microscopic mount.

The trama assumes four basic forms; regular, irregular bilateral and inverse.

The hymenium is a spore-producing layer composed either of fertile basidia only, or basidia and sterile, multiform cells.

These are cystidia of various types which are longer than basidia of various shapes, with a bare or encrusted surface, thin- or thick-walled, often growing out of the deeper layers of the gill tissue.

The cystidia situated on the gill edge, the so called cheilocystidia, are frequently arranged into continuos rows, sometimes different in colour from the gill face; such an edge, if observed with a lense is finely lacerate or ciliate. In some cases the cheilocystidia are filled with coloured pigment.

The cystidia present at the gill face are termed pleurocystidia and may be different from cheilocystidia in shape.

Gloeocystidia are charaterized by a granular or light-refracting content somewhat oily in appearance.

Basidioles are sometime present among them, if their shape is different they are termed cystidoles.

While in Agricales and Boletales the shape of the basidia is in the main uniform, in Aphylophorales it is considerably diversified and plays an important part in their taxonomic classification.

At the basidial apex there are usually four apicules each of which bears a single spore, basidia with two sterigmata are relatively common, while those having only one or alternatively having more than four sterigmata are rather rare.

Developmentally lower types of Basidiomycetes have septate basidia or basidia separating into two branches siderophilous basidia are typical of some genera of, Tricholomataceae—Lyophyllum, Tephrocybe, etc. their plasma contains granules staining black when exposed to ferric acetocarmine.

Trama in the broader sense of the word is tissue of the entire fruit body, whereas in the narrower sense it implies

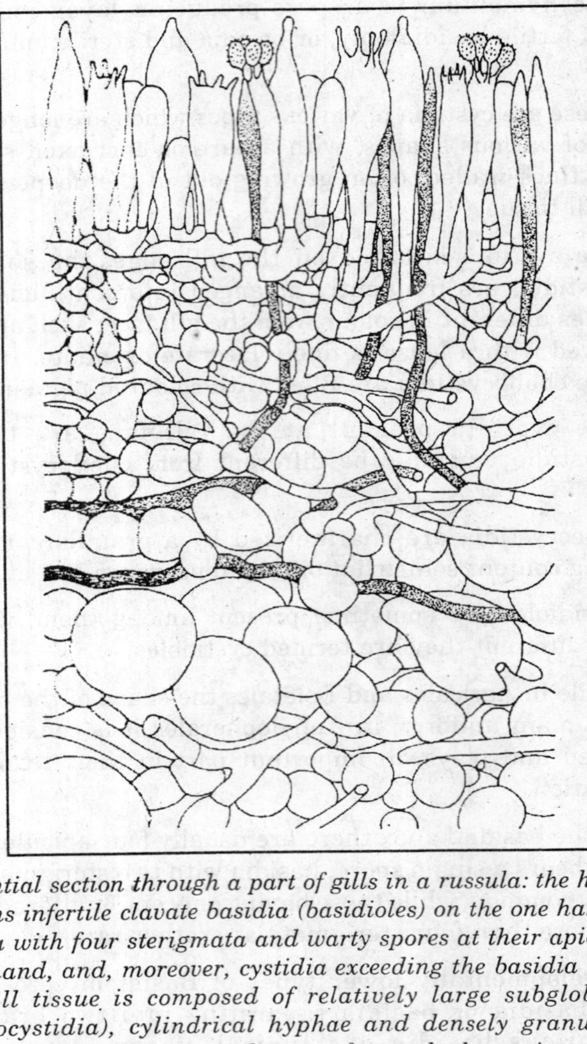

Tangential section through a part of gills in a russula: the hymenium contains infertile clavate basidia (basidioles) on the one hand, fertile basidia with four sterigmata and warty spores at their apices on the other hand, and, moreover, cystidia exceeding the basidia in length. The gill tissue is composed of relatively large subglobose cells (spherocystidia), cylindrical hyphae and densely granular, long lactifers extending in some places as far as the hymenium.

the gills, cap and stipe but excluding the surface tissues. The context itself, the fruit-body trama, is formed by a system of hyphae of the following three types: generative, binding and skeletal. Most common are fruit bodies consisting of generative (so-called basic) hyphae only; they are ramified, septate, and with or without clamp connections—this is known as the monomitic system. If, besides these, another type of hypae is present—either binding hyphae (which are more or less thick-walled, strongly fruticose, aseptate and lacking clamp connections), or skeletal hyphae (these are thick-walled, unbranched, long, aseptate)—this is known as the dimitic system. If all the three hyphal types are present in the tissue, they constitute the trimitic system. The fruit-body tissue of some fungus genera contains lactiferous tubes. These are very long, sinuous, usually aseptate hyphae filled with thick plasma of a frequently lacteal character, such as latex in the *Lactarius* species.

The context of fungi of the genera *Lactarius* and *Russula* contains—besides threadlike filaments and lactiferous tubes—clusters of globular cells, the so-called spherocystidia, conditioning the peculiar pomaceous brittleness of the flesh of these fungi. Of particular importance among the surface tissues is the cap cuticle which consists of one or more layers. The outermost layer of a multistratous cuticle is the epicutis, the layer between the epicutis and the cap context is known as the hypoderm or subcutis. If the cuticle is made up of clavate to globose cells, it is known as the hymeniform cuticle.

The colouring of fruit bodies depends upon physical processes taking place within the fruit bodies on the one hand, and upon the presence of pigments (whose distribution also has taxonomic significance) on the other. Intracellular pigments are present either in the plasma or in vacuoles, pigments are also in hyphal walls and in intercellubar spaces. The white colour is caused by the air filling intercellular spaces; if these are filled with water, the white colour becomes glassy and colourless (hyaline).

Spores are taxonomically the most relevant and constant characteristics. They are extremely multiform, uni- or multicellular, colourless or exhibiting a whole scale of colours, up to black. To be examined in detail, spores should always be observed under an immersion lens, at first in a colourless medium in water, in a 10-per cent solution of potash lye (KOH), in ammonia, etc.; another mount is coloured by

Melzer's reagent, cotton blue, or another colouring matter. Details not always clearly discernible in a cursory inspection of a microscopic mount are subjected to thorough examination: spore-wall thickness, presence of the apicule, shape and size of the germ pore, ornamentation of the spore wall, presence of the suprahilar disc or depression. One of the most significant discoveries that helped clarify the inter-relationships among gill fungi as well as among other fungus groups was the ascertainment of spore-wall amyloidity. Amyloidity is determined by Melzer's reagent with the aid of which the walls of spores or hyphae change either to some shade of blue (amyloid reaction), or to yellow-brown, red-brown or purple-brown hues (dextrinoid reaction). A more recent discovery is cyanophilia: here the spore or hyphal walls turn blue when touched with cotton-blue solution.

Only a brief mention will be made here of the disc fungi, whose larger representatives can often be come across by a mushroom-picker. The spore-bearing layer of their fruit body (apothecium) is composed of asci and paraphyses. The asci are cylindrical or clavate in shape and, as a rule, contain eight spores (ascospores) which are released either through pore in the tip of the ascus, or by the opening of a lid. In some genera the ascus walls are amyloid. Paraphyses, i.e. sterile filaments occurring among the asci, are often pigmented and consequently affect the colouring of the thecium (hymenium). In general, what has been said about spores of the basidiomycetes applies also to ascospores. The context of the apothecium—which is either directly attached to the substrate or seated on a sterile stem or stipe—has a relatively simple structure and its external layer bears various growths, such a bristles or hairs.

4

Reproduction of Fungi

In Fungi reproduction takes place in two ways.

1. Sexual reproduction
2. Asexual reproduction.

Both of them are commonly encountered. Sexual reproduction is generally regarded as a union of two sexually differentiated nuclei, e.g. in spores.

The germinating spores form a unisexual mycelium, consisting of hyphae with uninucleate cells. Each of these cells contains a nucleus with a certain number of chromosomes.

The primary mycelium may go on growing for a very long time without forming any sex organs. If two uninucleate without forming any sex organs.

If two uninucleate sexually differentiated mycelia meet the plasmas of the conjugating cells, unite into a single binucleate cell the dicaryon.

The dicaryon grows into hyphae made up of binucleate cells into the secondary mycelium. Even this mycelium can take a very long time to grow; under suitable external conditions, however it gives rise to fruit bodies. Hence it follows that fruit bodies are composed of hyphae with binucleate cells.

The fusion of the two dicaryotic nuclei takes place only at the hyphal end in terminal cells transformed into basidia or asci.

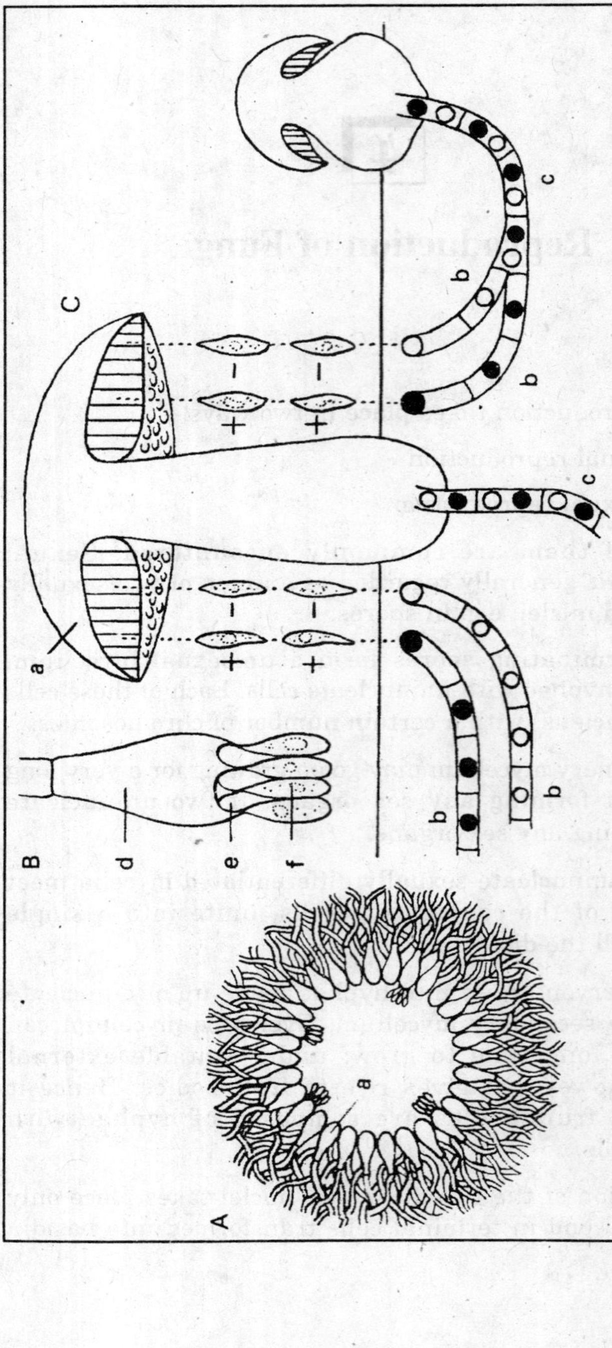

Reproduction in Basidiomycetes:

(A) corss section through a tube of a bolete; (B) formation of spores on a basidium; (C) development of a fruit body from the mycelium; (a) hymenium; (b) primary mycelium; (c) secondary mycelium; (d) basidium; (e) sterigma; (f) sexually differentiated spores.

The single nucleus arisen through the association of two dicaryotic nuclei has a double number of chromosomes (2n) that is a double amount of nuclear matter and thus of hereditary information. It is only through the subsequently reduction division that the number of chromosomes is lowered to the original state (n). These nuclei give rise to spores—basidiospores on basidia ascospores in asci. Even though, of course there are exceptions to this scheme, it is sufficient for basic understanding of the principle of sexual reproduction in fungi.

A sexual reproduction is never preceded by a fusion of the nuclei of two cells. Its simplest form involves the decomposition of the mycelium into simple fragments termed arthrospores. Another form is the abstraction of independent germ spores (conidi) from specialized hyphae (conidiophores).

The conidia vary is shape, size and colouring; they may be unicellular or multicellular. Chlamydospores are thick-walled spores developed from mycelial cells. Their function is that of resting cells capable of surviving long-lasting unfavourable living conditions.

Cells giving rise to asexual spores are generally designated as conidiogenous.

In classifying the conidial stages of fungi a great many of which occur independently and are included in the subsidiary class of imperfect fungi (Deuteromycetes, formerly also Fungi imperfecti).

Particular importance is attributed to the manner in which asexual spores are formed.

Spores originating in any of these ways are produced by fungi in such enormous quantities that they not only populate the whole biosphere of our planet but can be found even in the atmosphere. As spores the fungi are omnipresent, and most of them seems to be endowed with the capacity to withstand unfavourable influences and to remain in a latent stage without losing their viability.

This over production of spores is a necessary pre-condition for the survival of the species.

5

Basic Mycological Terms

To understand, more things thoroughly it is very much needed to learn important mycological terms. Now let us try to learn more about these terms.

cleistothecium – a closed globulor fruit body of the Ascomyceles, opening by rupture.

epigeous – developing above the surface of the ground.

excipulum – the external tissue forming the apothecium.

fimicolous – growing on dung or excrements.

collarium – a collar like arrangement of gills about the apex of the stipe.

columella – a sterile part of the sporangium or fruit body usually leading from the apex of the stipe or from the base into the spore-bearing part (in slime fungi moulds and some gasteromycetes).

conidiogenous cell – a cell bearing conidea.

fructification – the stage or period when the fungus produces fruit bodies.

fruit body – a part of the thallus differentiated in shape, consisting of hyphae and reproductive spore producing cells.

conidiophore – a specialized independent hypha either simple or branched bearing conidiogenous cells and conidia.

gametangium – an organ producing gamets, i.e. sexual-reproduction cells.

autotropic nutrition – the capacity of the plants to transform absorbed inorganic substances into organic ones.

acyanophilous – not staining blue when touched with cotton blue.

aethalium – a pulvinate or semiglobular fruit body of some slime moulds arising from the association of a large number of sporangia.

allantoid spores – curved cylindrical spores, sometimes also referred to as sausage-shaped.

amyloid – spore walls or hyphae staining blue in solutions containing iodine. In mycology, Melzer's reagent is used for testing amyloidity; the colouration appears in various shades of blue (from bluish grey over deep blue or bluish green to black).

anamorph – a stage of the fungal thallus which is either sterile (e.g. sclerotium, rhizomorph), or producing asexual spores (conidia, chlamydospores). Most of the conidial stages in fungi are anamorphs.

annulus – ring; in general it is a remnant of the veil (*velum partiale*) on the fruit-body stipe, usually in the form of a membranous collar.

anthracophilous – inhabiting burned-over soil or growing directly on burned wood.

apical – at the apex.

apiculus – a minute projection at the basal end of the spore by which the spore is attached to the sterigma.

apothecium – the fruit body of the Discomycetes with an open thecium.

ascocarp – the pseudoperithecium of some Pyrenomycetes.

ascospore – a spore of the Ascomycetes borne in an ascus.

ascus – a saclike, cylindrical or claviform structure in the Ascomycetes where, after nuclear association and a subsequent reduction division, spores (ascospores) develop.

asexual reproduction – vegetative reproduction; in general, this implies reproduction through elements arisen without any foregoing fusion of the plasmatic and nuclear contents of two sexually different cells.

assimilative pigment – photosynthetic pigment (particularly chlorophyll in green plants).

basidiole – a sterile basidium (without sterigmata) or other sterile hymenial cells resembling the basidia in both shape and size.

basidiospore – a spore of the Basidiomycetes separated from the basidium by a sterigma (pedicel).

basidium – a cell mostly clavate or cylindrical in shape, bearing pedicellate (sterigmate) spores (basidiospores) developing subsequent to nuclear association and the reduction division following thereon. It is characteristic of the Basidiomycetes.

bilateral gill trama – see trama.

binding hyphae – mostly thick-walled, richly branched fruticose hyphae with a short basic pedicel, aseptate, lacking clamp connections.

bitunicate ascus – a double-walled ascus.

capillitium – sterile fibres mixed with the spores within the fruit bodies of some Gasteromycetes or Myxomycetes.

carminophilous basidia – see siderophilous basidia.

caulocystidium – a cystidium developed on the stipe cuticle.

cheilocystidium – a cystidium occurring on the edge of a gill.

chemical reagents – are used in identifying fungi as well as in preparing microscopic mounts. The most common ones are a 3-10 per cent aqueous solution of ammonia, a 2-5 per cent solution of potassium hydroxide (KOH), lactophenol, lactic acid, Melzer's reagent, cotton blue, acetocarmine. Colour reactions of the context in fresh fruit bodies are tested e.g. by an aqueous green-vitriol solution, phenol, sulphovanillin, benzidine, etc.

chlamydospore – a thick-walled vegetative spore developed from a hyphal cell.

chrysocystidium – a claviform cystidium in gill-bearing fungi filled with an amorphous, often yellow-coloured matter.

clamp connection – a protuberance arching over the hyphal septum in numerous Basidiomycetes.

conidium – an asexually (vegetatively) produced spore on a hypha or conidiophore.

coprophilous – inhabiting excrements or dung.

copulation – the fusion of two sexual cells, or of the contents of sexually different cells.

cortina (cobweb) – a membranous veil covering the hymenium only, eventually a part of the epicutis of the cap.

cotton blue (CB) – dissolved in lactic acid, it is used to colour hyphae, spore walls, and particularly the spore ornamentation, mostly in the Discomycetes.

cyanophilous – readily turning blue to violet in a cotton-blue solution; the opposite is acyanophilous.

cystidiole – a sterile cell in the hymenium situated between the basidia, arising from the same hymenial level as the basidium but differing from it in shape.

cystidium – a sterile cell in the hymenium of the Basidiomycetes; it is larger, different in shape from the basidium, and usually more deeply embedded in the hymenium.

dendrophysis – much branched, often dendriform cell found in the hymenium or in the cap cuticle.

dermatocystidium – a cystidium growing from the cuticle of the fruit body; the pileocystidium develops on the pileus cuticle, the caulocystidium on the stipe cuticle.

detritus – a layer of decomposed remains of organic matter (mostly of plant origin) lying on the surface of the ground.

dextrinoid – staining yellowish to reddish brown or vinaceous (purplish) brown in Melzer's reagent; it means the same as pseudoamyloid.

diploid – having a twofold number of nuclear chromosomes, as e.g. in the zygote.

eccentric stipe – attached to the pileus extracentrally but not laterally.

epicutis – the outermost layer of the multistratal cuticle covering the pileus of gill fungi; it differs from the other layers in structure.

generative hyphae – basic ramified, thin-walled, septate (partitioned) hyphae with or without clamp connections, filled with plasma (see hyphal systems).

germ pore – a usually flattened spot in the thinned part of the spore wall through which the spore germinates.

gill folds – the hymenophore assumes the form of thick and narrow ribs covered with a continuous hymenial layer.

gleba – the fruit-body tissue of the Gasteromycetes producing basidia.

gloeocystidium – a thin-walled cystidium generally having a refractive or granular (often oily) content. It is elongate, claviform or cylindrical, usually deeply anchored in the hymenium, sometimes also in the trama or cap cuticle.

haploid – having half the number of chromosomes in the nuclei of germ cells (gametes).

heterotrophic organisms – having no assimilative pigments, they derive organic substances directly from the living or dead tissues of autotrophic organisms (green plants) or animals.

hilar depression – a slightly depressed spot on the ventral spore wall right above the apiculus.

host – a plant or animal with which either the entire life cycle of a parasite or saprophyte, or at least a part of it is connected.

hydnoid – having a spinulose hymenophore.

hygrophanous – the cap in gill fungi is darker coloured when damp than after drying out. With the moisture disappearing it fades in colour and is often divided into two contrasting colour zones (the apical and the marginal), or is radially striped with darker and lighter colour shades.

hymeniderm – the structure of the hymenidermal cuticle of the cap in similar to that of the hymenium (all cells are arranged on the same level).

hymeniform – a layer of globular, pyriform, shortly cylindrical or claviform cells arranged in a palisade (e.g. the cap cuticle in some gill fungi).

hymenium – a layer of basidia arranged palisade-like.

hymenophore – part of the fruit body bearing the hymenium.

hypha – filament; the basic structural unit of the fungus body, usually divided by septae into longer and shorter cells; hyphae constitute the mycelium and the major part to the fruit body.

hyphal systems – the fruit body can be composed of one, two or three types of hyphae: the monomitic system consists exclusively of basic (generative) ramose hyphae with or without clamp connections; the dimitic system is made up of generative and skeletal, or generative and binding hyphae; the trimitic system has generative, skeletal and binding hyphae.

hypogeous – growing underground.

hypothallus – also subiculum; a mass of intertwined surface hyphae forming a felted or cobweb-like layer from which fruit bodies arise.

hypothecium – a layer of tissue immediately beneath the thecium of the Discomycetes.

hysterothecium – the usually elongate fruit body of Ascomycetes or lichens with a narrow longitudinal slit in the middle.

incrustation – the surface of hyphal walls is coated with a hard crust, granules or small crystals of frequently refractive colourless matter (incrusted hyphae).

irpicoid – the hymenophore assumes the shape of irregular, flattened (foliated) teeth or short lamellae.

irregular – see (gill) trama.

jellification – coagulation into jelly; the walls of jellified hyphae are coated with a layer of usually colourless jelly.

lactifers (laticifers) – long, nonsepatate hyphae bearing thick, granular plasma or latex (milky-coloured juice).

lamellae – gill-like structures of the hymenophore radially spreading out from the stipe to the cap margin

lamellules – short gills beginning in the cap margin but not reaching the stipe.

lamprocystidium – a thick-walled cystidium in gill fungi.

lateral stipe – attached to one side of the cap.

lignicolous – growing on wood.

macromycetes – an auxiliary term applied to large fungi whose fruit bodies are visible with the naked eye (they usually exceed 0.5 cm in size).

margo – marginal edge of the apothecium.

meiosis – one of the two successive nuclear divisions of a diploid nucleus is reductional, the result being four daughter nuclei having a haploid (n) number of chromosomes.

Melzer's reagent – the solution of chloral hydrate in water with an admixture of potassium iodide (KI) and iodine (I) used for testing amyloidity (KI, 1.5 g; I, 0.5 g; distilled water, 20 cc; chloral hydrate, 22 g). It was discovered and used for the first time by the Czech mycologist V. Melzer and is one of the most generally applied and most important reagents in taxonomic study and in identifying fungi.

metachromatic – a layer of cells or hyphae changes colour through the application of a given colouring agent, yet the newly acquired colour differs from that originally borne by the colouring agent – e.g. cresylic blue discolours to red.

metuloid – a thick-walled cystidium with a strongly incrusted surface arising in the deep layers of the hymenophore tissue.

micromycetes – an artificial grouping of fungi of predominantly microscopic dimensions whose fruit bodies usually do not surpass the limit of discernibility with the unaided eye. This group includes all imperfect fungi, rusts, smuts, moulds, and the greater part of Ascomycetes.

micron – 1 μm = one thousandth of a millimetre. In mycology, all microelements subjected to microscopic study are indicated in microns.

monomitic – see hyphal systems.

mycelium – the vegetative part of the fungal thallus obtaining nutritional substances from the substrate.

mycology – the science of fungi.

mycorrhiza – a symbiotic association of fungus hyphae with the roots of green plants.

nonamyloid – failing to turn blue in Melzer's reagent.

ocrea – partly membranous remnants of the universal veil (*velum generale*), attached to the base of the stipe like a sheath, having a distinct upper margin. They are typical of the *Cortinarius* species belonging to the subgenus *Telamonia*.

oidium – a thin-walled cell arising from the disintegration of a vegetative hypha.

oogonium – a female sexual organ.

oosphere – an egg cell.

oospore – a thick-walled spore formed as the result of fertilization of the oosphere.

ornamentation of spores – the spore-wall surface is sculptured with various excrescences (warts, spinules, ridges, reticulation).

ostiole (ostiolum) – a part of the perithecium narrowing down into a neck-shaped extension provided with a canal lined with hyphae and opening by a pore.

palisade tissue – more or less parallel claviform or cylindrical elements usually situated on the pileus surface or composing the hymenium.

paraphyses – sterile filaments filling up the space between asci (see ascus).

parasite – an organism living or growing upon other living organisms from which it derives nourishment and, in doing so, damages or destroys them.

peridium – the cover enveloping the fruit body in Gasteromycetes; outer peridium = exoperidium, inner peridium = endoperidium.

perithecium – a closed, usually more or less rounded fruit body characteristic of the Pyrenomycetes, provided with a wall which, as a rule, opens at the apex by an ostiole.

phialide – a specific cell on the conidiophore, giving rise to conidia.

pileocystidia – cystidia situated on the pileus surface.

plage – suprahilar disc; a more or less circumscribed, smooth spot on the spore wall immediately above the apiculus.

plasmodiocarp – the extending, sessile and ramose fruit body of slime moulds.

plasmodium – a body of plurinucleated, naked protoplasm in Myxomycetes exhibiting slow motion and ultimately changing into fruit bodies.

pleurocystidium – a cystidium occurring on the face of gills.

pore – an opening at the apex of the ascus, or an opening in the ostiole of the perithecium.

poroid – a hymenophore consisting of tubes.

pseudoamyloid – see dextrinoid.

pseudoparaphyses – paraphysoid filaments; sterile threads in some Pyrenomycetes separating the single asci contained in the pseudothecium.

pseudoparenchyma – fungous tissue composed of globular or subglobular cells.

pseudothecium – the perithecium-like fruit body of some Pyrenomycetes; developmentally, however, it is a stroma containing bitunicate asic and pseudoparaphyses – i.e. remnants of the inner tissue.

pycnidium – a closed, mostly ball-shaped fruit body producing nothing but conidia.

radial – spreading out radially from the centre to the margin (for example, fibres on the pileus surface).

receptaculum – an alternative designation for the apothecium or for its greater part; also the bearer in Phallales.

resupinate – reclining over the surface and mostly attached to the substrate (fruit bodies of numerous Aphyllophorales).

rhizoid – a rootlike or strandlike mycelium at the base of the fruit body.

rhizomorph – mycelial strands with a thickened (sclerotized) outer layer.

ring – see annulus.

saprophyte – living on dead organisms, their remains, and on organic matter of mineral soils.

sclerotium – a tuberiform, lenticular or irregular structure composed of firmly intertwined hyphae; it is a resting stage from which, under certain conditions, fruit bodies or hyphae may develop.

sections through the gills – the tangential section is made across the gills in the marginal part of the cap;

– the longitudinal (radial) section is cut through the centre of the cap so that it mostly passes between the gills or longitudinally bisects them;

– the cross (transverse) section is cut horizontally to the surface of the ground (in fruit bodies growing on the ground).

septum – partition; a transverse or longitudinal wall in fungus hyphae and spores.

seta – a thick-walled, tough, usually dark-coloured cystidium tapering upward, or a structure similar to it.

sexual reproduction – fusion of the plasmatic and nuclear content of two sexually differentiated cells, and the subsequent development of spores.

spherocystidium – a globular cell in the tissue of Basidiomycetes (e.g. in the genus *Russula*).

siderophilous basidia – their plasma contains a large quantity of granules turning dark in ferric acetocarmine; the same as carminophilous basidia.

skeletal hyphae – thick-walled, unbranched, long, nonseptate hyphae; see also hyphal system.

sporangiophore – a hyphal branch bearing the sporangium.

sporangium – an asexually produced cell bearing one or more spores (e.g. mycetozoan fruit bodies).

spore – the reproductive monocellular or multicellular body functionally corresponding to the seed in higher plants but lacking an embryo.

sterigmà – the minute pedicel on the surface of the basidium upon which a spore is borne.

stroma – a rigid structure composed of hyphae wherein fruit bodies or cavities containing asci or conidiophores are embedded (mostly in Pyrenomycetes).

subhymenium – a layer of tissue under the hymenium (is structure in different from that of the trama).

subiclum – see hypothallus.

substratum – the nutritive base (e.g. earth, soil, plant remains, wood, excrements).

suprahilar dise – see plage.

terrestrial – growing on the ground.

thecium – the uppermost part of the Discomycetes fruit body with the hymenium; the latter consists of asci and paraphyses arranged in a palisade.

thallus – a simple plant or fungus body undifferentiated in roots, stalks or leaves (gills).

trama – the tissue supporting the hymenium in Basidiomycetes; in gill fungi, the following four types of gills are distinguished: regular trama (the hyphae are parallel to each other); irregular trama (some hyphae diverging in various directions are interwoven with the parallel hyphae); bilateral trama (the hyphae are parallel only in the mediostratum of the gills, while in the margins they diverge towards the hymenium in curve); inverse trama (similar to the bilateral trama but the hyphae are diverging backwards.)

trichoderm – the cuticle consisting of densely clustered and erect hairs.

trimitic – see hyphal systems.

tubes (tubules) – vertically arranged tubular structure composing the hymenophore.

vegetative reproduction – see asexual reproduction.

velum (veil) – the cover enveloping the whole fruit body or a part of it; *velum universale (generale),* universal veil, enfolds the entire fruit body; *velum partiale,* partial veil, connects the cap margin to the apex of the stipe, covering the hymenophore when young.

volva (cup, sheath) – a usually membranous, cup-shaped structure around the base of the stipe (a remnant of the universal veil).

zygospore – a resting spore resulting from the conjugation of the contents of two gametangia or two gametes.

zygote – a cell in which two nuclei of the opposite sex have fused together. It is diploid, having a double number of chromosomes.

6

The Classification of Fungi

The classification of fungi differs according to individual authors. So we can say the views about the classification of fungi one far from uniform.

In this chapter the classification is restricted to merely a basic survey of the orders constituting the two principal groups, of the so called true fungi. That is Ascomycetes and Basidiomycetes.

Mycetozoans now-a-days generally regarded as fungi and also moulds (Phycomycetes) remain outside the sphere of our interest.

Now let us study about the, Phylum, class, subclass, and order of Fungi.

Fungi

Phylum: true fungi — Eumycophta

Class: Ascomycetes.

Subclass: Protoascomycetidae.

Order: Eurotiales.

Order: Microascales.

Order: Onygenales.

Order: Laboul beniales.

Subclass: Ascohymenomycetidae.

Order: Erysiphales.

Order: Pezizales.

Order: Tuberales.

Order: Helotiales.

Order: Phacidiales

Order: Xylariales.

Order: Hypocreales.

Subclass: Ascoloculomycetidae.

Class: Basidiomycetes.

Subclass: Heterobasidiomycetidae.

Order: Protoclavariales.

Order: Auriculariales.

Order: Tremellales.

Order: Uradinales.

Subclass: Holobasidiomycetidae.

Order: Aphyllophorales.

Order: Polyporales.

Order: Boletales.

Order: Agaricales.

Order: Russulates

Class: Gasteromycetes.

Auxiliary Class: Deuteromycetes.

Ascomycetes

Ascomycetes are usually regarded—together with Basidiomycetes—as higher fungi. The common feature of all, Ascomycetes is the production of spores in spore sacs (asci) most often eight to an ascus. The micelium is always septate, and the septa are, furnished with a simple minute opening facilitating the passage of, plasma and nuclei.

The principal constituent of, mycelial walls, as well as all hyphal walls, in the fruit body is chitin.

Major part of Ascomycetes, produce fruit bodies on their mycelium which basically assume, two forms substantiating the division of the fungi into two main groups.

Group I

The fruit body is usually globular, closed, opening at, maturity either by rupture (the cleistothecium) or by a small pore at the apex.

Group II

The fruit body is freely open either from the very beginning or at least in the later stages. Together with the asci, sterile filaments (paraphyses) also develop within the fruit body. Many species also form asexual stages to which the designation anamorphs is now-a-days applied, while the ascus stages are termed teleomorphs.

The large group of fungi, Ascomycetes includes about 40,000 hitherto described species. Their fruit bodies usually no more than 1 mm in size occurring in nature on most various substrates.

Now let us study about, the subclass, proto-uscomycedidae, which have globular asci with a thin, simple wall provided with no opening apparatus.

Fungi of the order **Eurotiales** are saprophytes currently distributed in soil and humus where they decompose organic remains. The so-called moulds of the genera *Penicillium* and *Aspergillus* belong here – of course both these names relate only to the conidial stages, valid names for the much rarer asucs stage of the genus *Penicillium* being *Talaromyces,* for *Aspergillus, Eurotium.* This order, however, also includes some larger fungi with globular fruit bodies up to several centimetres in size. They are represented in our forests by the genus *Elaphomyces.*

The order **Microascales** has become notorious particularly by its microscopic member *Ceratocystis ulmi* whose

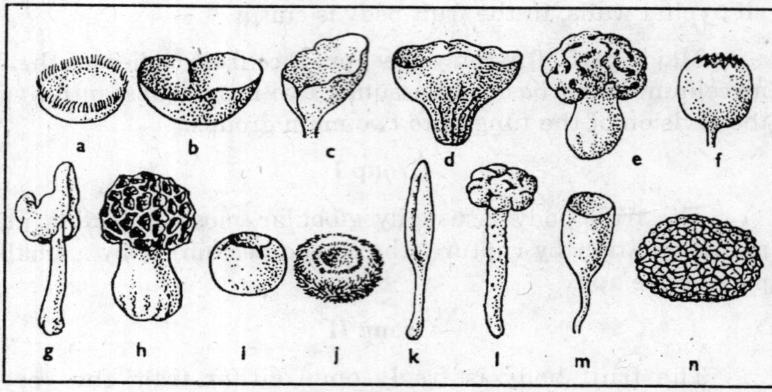

Examples of fruit bodies in Ascomycetes: (a) Scutellinia; (b) Peziza; (c) Otidea; (d) Helvella; (e) Gyromitra; (f) Tarzetta; (g) Helvella elastica; (h) Morchella esculenta; (i) Peziza vesiculosa; (j) Mycolachnea hemisphaerica; (k) Geoglossum; (l) Leotia lubrica; (m) Urnula craterium; (n) Tuber aestivum.

mass occurrence has led to the extermination of most European elms. The disease caused by this fungus is called Dutch elm disease.

Fungi of the order **Onygenales** have small, stipitate, globular fruit bodies growing on the horny tissue of claws, hoofs, horns, but also on the feathers of dead birds. These ceratophilous fungi are represented by the genus *Onygena*.

Fungi of the order **Laboulbeniales** are sometimes regarded as an independent class of Ascomycetes and their position in the system is still ambiguous. They are narrowly specialized parasites of insects (especially beetles) and spiders to which, however, they do not seem to cause much harm. Their minute, fusiform fruit bodies are fastened to the surface of the host's body by a 'foot'.

Fungi of the subclass **Ascohymenomycetidae** have asci whose length usually exceeds their breadth. Their apex is equipped with an opening apparatus or a lid for discharging ascospores. The ascus wall is thin, and does not turn slimy when mature. They are usually divided into seven orders:

The order **Erysiphales** (mildews) includes the most common and also most conspicuous parasites of vascular plants. Their surface mycelium often forms extensive coatings, particularly on the leaves, which are either white, farinaceous (family Erysiphaceae), or brown to blackish (Meliolaceae, currently distributed almost exclusively in the tropics or subtropics). Minute, globular, closed fruit bodies (cleistothecia), often provided with variously shaped filaments, appear to the naked eye as dark little dots. Numerous pests of cultivated plants belong to this group, for example *Uncinula necator,* grapevine mildew (known especially in its conidial stage as *Oidium tuckeri*), further *Podosphaera leucotricha* on apple-trees, *Microsphaera quercina* on oaks, etc. Traditionally the order Erysiphales was classed with the subclass Protoascomycetidae referred to above; at present, however, it has been reclassified as a member of the subclass Ascohymenomycetidae, with respect to its asci opening through a lid.

The following four orders are currently referred to as cup fungi (Discomycetes).

The fruit body of **Pezizales** is a typical apothecium, globularly closed only in youth; soon it opens into its definitive shape which is semiglobular, patelliform or discoid, often flatly expanded or even convex. The inner surface, known as the disc (thecium), bears asci and paraphyses arranged in palisade-like structures. Fruit bodies in the most developed forms, Helvellaceae and Morchellaceae, are differentiated into a stipe and a cap. Asci open by a lid and ascospores are shot out. In some genera the ascus membranes are markedly amyloid. As well as numerous representatives of minute size, Pezizales also include the largest and hence the most conspicuous species of Discomycetes. Some of them show remarkable colours, especially red, orange and yellow. They grow almost exclusively as saprophytes on bare ground, in moss, on putrescent wood or on other plant remnants, many of them on excrements, dung and on burned-over soils. They are generally divided into six families. An important identifying feature is – besides other characteristics – the ornamentation of their ascospores. This book deals with species of the families Humariaceae, Otideaceae, Pezizaceae, Helvellaceae and Morchellaceae.

The order **Tuberales** consists of hypogeous fungi. The tuberiform fruit body remains closed, the incidence of spore-bearing locules opening on its surface is rather rare. Asci usually cover the walls of galleries and cavities inside the fruit body. Spores are often large, ornamented; unlike the preceding species, they are not ejected out of the asci but get free through the decomposition of the entire fruit body. The Tuberales are mycorrhizal fungi distributed mainly in warm regions, in deciduous forests, and on calcareous soils. They include the families Tuberaceae and Terfeziaceae.

The order **Helotiales** is the largest group of cup fungi. Their lidless asci open at the apex by a ring-shaped apparatus through the centre of which the spores are violently discharged. Their fruit bodies are typical apothecia, mostly minute in size (1 mm and less in diameter), sessile or stipitate, smooth or hairy, bristly and felted, in some genera embedded in the substrate at first and pushing their way to the surface; more advanced types have upright fruit bodies with long stipes, cylindrical or pileate (e.g. in the family Geoglossaceae). They grow predominantly as saprophytes, rarely as parasites, on plant stems, leaves, wood or rind.

The fruit body of fungi belonging to the order **Phacidiales** is called the pseudoapothecium; this is embedded in the tissue of the host plant and covered with a black stromatic layer which bursts open at maturity, disclosing the disc. The disc is covered up in dry conditions. The fruit body opening through a longitudinal slit is called the hysterothecium and is characteristic of the family Hypodermataceae. This includes for example the genus *Lophodermium;* some of its species live as needle parasites, causing a disease known as needle cast. Other typical representatives of this order are the genus *Rhytisma* whose conidial stage causes black spottiness on maple leave, and *Colpoma* with the species *C. quercinum,* commonly found on dead oak branches.

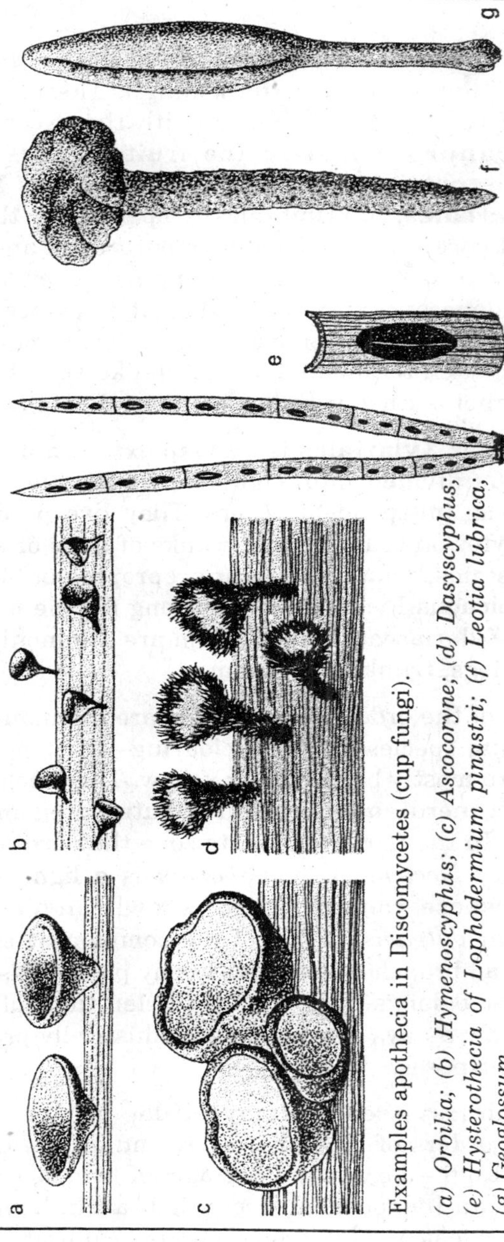

Examples apothecia in Discomycetes (cup fungi).

(*a*) *Orbilia*; (*b*) *Hymenoscyphus*; (*c*) *Ascocoryne*; (*d*) *Dasyscyphus*; (*e*) *Hysterothecia* of *Lophodermium pinastri*; (*f*) *Leotia lubrica*; (*g*) *Geoglossum*.

The last-named two orders have been give the common designation Pyrenomycetes (flask fungi). These include also the subclass Ascoloculomycetidae (with the exception of the order **Lecanorales**). Here the fruit body is a typical perithecium containing asci which, in the orders **Xylariales** and **Hypocreales**, are unitunicate, opening at the apex by an amyloid pore or by an opening apparatus of another kind. The globular or lageniform perithecium has its own wall made up of pseudoparenchyma, even when it is embedded in the stroma. It ordinarily opens through a minute circular opening at the apex, often terminating in a neck-like extension (ostiole) whose channel is lined with hyphae (periphyses).

The order **Xylariales** is a group extraordinarily rich in species; it is divided into a number of families sometimes considered as independent orders. They live predominantly as saprophytes on branches and trunks of trees or shrubs and on herbal stems; many species are coprophilous. Larger and more conspicuous Pyrenomycetes belong for the most part to the family Xylariaceae; some of them are commonly found on dead wood logs, trunks and stumps.

Fungi of the order **Hypocreales** are distinguished from the preceding species by their colouring—their perithecia or stromata are mostly bright red or yellow. They include a large number of genera and species distributed predominantly in the tropics. In the north temperate zone they are represented by the genus *Nectria*. *N. cinnabarina* is a lignicolous pyrenomycete whose conidial stage, known under the name *Tubercularia vulgaris,* is one of the commonest small fungi living on dead branches of various woody plants. Also the well-known ergot belongs here; it is the sclerotium of *Claviceps purpurea,* now-a-days cultivated and industrially processed for medicinal purposes.

The subclass **Ascoloculomycetidae** includes on the one hand three orders of Pyrenomycetes, and on the other hand lichenized fungi – i.e. the greater part of lichens subjected to research by an independent scientific branch, lichenology. A characteristic feature of this subclass are bitunicate asci whose walls consist of two layers. After the spores have attained maturity, the outer layer bursts open at the apex, while the inner layer juts out of the newly created opening and the ascospores are discharged. The small fruit bodies resemble the perithecium but are regarded as a special structure termed the ascocarp (also pseudothecium or ascostroma). It is in fact a stroma consisting of one or more locules with no special wall dividing them from each other. This group comprises a vast amount of genera and species of both parasites and saprophytes inhabiting most various parts of plants. No uniform classification of orders and families has so far been worked out.

Basidiomycetes

This is the developmentally most advanced class of fungi, including at least 20,000 species. The characteristic feature common to all fungi of this group lies in the origin of spores (basidiospores): they develop on specialized cells called basidia. A young basidium has two nuclei coalescing into a single nucleus. A twofold reduction division gives rise to four haploid daughter nuclei entering by apical or lateral pedicels (sterigmata) into the basidiospores borne at the tips of sterigmata. Basidiomycetes of a less advanced type usually possess divided or septate basidia. The shape of basidia remains constant and is considered of paramount taxonomic significance in determining mutual relationships. Basidiospores are unicellular and are constant in shape and size in the individual taxons. Another important distinguishing feature is their wall, often ornamented with various excrescences, and amyloid in many species or genera. Basidia are usually aggregated in a continuous layer (hymenium), wherein special sterile cells (paraphyses, cystidia) are often present. The mycelium consists of simple or branched hyphae penetrating through the substrate; sometimes they are quite conspicuous. Mycelial hyphae are septate, consisting mostly of chitin. Each septum is furnished with a 'dolipore' (an opening in the central, bilaterally thickened part of the septum covered by a semiglobular membrane). In the primitive types, the fruit body itself is reduced to a layer of hyphae bearing the hymenium or single basidia only; they usually assume the form of coatings or membranes which are termed resupinate fruit bodies. More highly developed forms usually possess an upright fruit body differentiated into the pileate part (cap, pileus) and the stipitate part (stem, stipes). The hymenium, which in the simpler forms covers the entire outer surface of the fruit body, is confined in the more advanced forms to a specific, exactly delimited part, the so-called hymenophore, situated usually on the underside of the cap. The surface of the hymenophore is often enlarged by various excrescences, tubes or gills. The fruit body either has no apparent cover (a gymnocarpous fruit body) or is enclosed – at least in the initial stages of development – in a special covering (veil, velum). The spore-producing tissue sometimes develops within the fruit body, as for instance in the Gasteromycetes.

Basidiomycetes usually live as saprophytes on decaying plant remains, on the forest floor or in humus and duff; many of them form symbiotic relationships with roots of green plants (mycorrhiza). Most species are confined to the forest environment: here they play a significant role and their fruit bodies represent a remarkable phenomenon especially in the summer and autumnal aspect of plant communities. Their principal function consists in their contribution to humus formation, to the decomposition of wood – nor is it possible to overlook their role as a means of sustenance.

The shape of basidia underlies their division into two subclasses – Heterobasidiomycetidae and Holobasi-dio-mycetidae.

The subclass **Heterobasidiomycetidae** includes Basidiomycetes with either simple or – most frequently – septate (i.e. multicellular) or more or less deeply divided basidia. It is formed by the following four orders:

Fungi of the order **Protoclavariales** have basidia consisting of a globular or cylindrical part (hypobasidia) to which inflated or elongate epibasidia are attached, bearing sterigmata with spores. The fruit bodies are often remarkably gelatinous or elastic. These small fungi usually live as saprophytes on dead wood. They are classed into two families, Tulasnellaceae and Dacrymycetaceae; a common representative of the last-mentioned family, *Calocera,* is dealt with in this book.

Fungi which belongs to Auriculariales, usually have cylindrical or clavate basidia divided into several cells by transverse septae.

A single sterigma ending in a spore develops from each of these cells.

Fungi of the order, Tremellales produce almost globular cruciately septate basidia, divided by one or two longitudinal septae perpendicular to each other into two or four parts. The fruit bodies are mostly gelatinous, and are of various shapes into two or four parts. The fruit bodies are mostly gelatinous and are of various shapes. This order includes predominantly wood saprophytes, of the larger ones particularly various species of the genus *Tremella* and *Pseudohydnum* belonging to the family Tremellaceae.

The order **Uredinales** (rusts) includes micromycetes parasitizing green plants in which they cause the so-called rustiness; the attacked plant is covered with minute rusty, orange, brown to black-coloured deposits in the form of tubercles or other structures. The tissue of the infested plants often withers, in other cases rusts may cause hypertrophic growth. Rusts that attack cultivated plants, especially cereals, are of considerable importance. Numerous rust species produce four types of spores: aeciospores (spring) in cupulate aecia, urediospores (summer) in uredia, teliospores (winter) in telia; teliospores germinate into the promycelium from which sporidia arise (the fourth type of spores). A characteristic of the rusts is an alternation of sexual and asexual generations; in some of them there is also an alternation of hosts on the one hand, and a narrow specialization on a certain (often a single) type of host on the other. The substantial part of species is concentrated in the genera *Puccinia* and *Uromyces*.

In the past, smuts (Ustilaginales) used to be introduced here as the following order. Today they are included in the

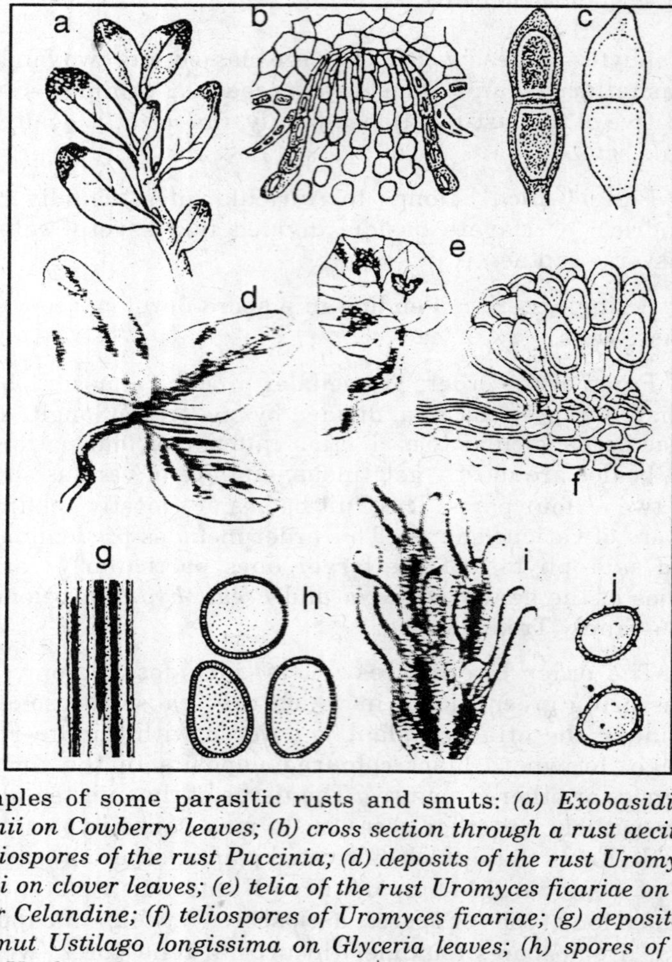

Examples of some parasitic rusts and smuts: *(a) Exobasidium vaccinii on Cowberry leaves; (b) cross section through a rust aecium; (c) teliospores of the rust Puccinia; (d) deposits of the rust Uromyces trifolii on clover leaves; (e) telia of the rust Uromyces ficariae on the Lesser Celandine; (f) teliospores of Uromyces ficariae; (g) deposits of the smut Ustilago longissima on Glyceria leaves; (h) spores of the smut Ustilago longissima; (i) cereal grains attacked by the smut Ustilago segetum; (j) spores of Ustilago segetum.*

class Endomycetes which represents the transition between moulds (Phycomycophytina) and primitive Ascomycetes. Other authors, however, join smuts and rusts into the independent class Teliomycetes. Smuts, like rusts, parasitize vascular plants.

The subclass **Holobasidiomycetidae** is the developmentally most advanced group of Basidiomycetes.

Fungi belonging to this subclass have unicellular basidia usually bearing four sterigmate spores at their apex. The hymenium is either freely expanded or enclosed in the fruit-body tissue. The subclass is divided into six orders. The first five are the so-called Hymenomycetes in which the hymenium is freely opened before the spores have reached maturity and covers either the entire fruit-body surface or some of its parts. The spores are actively discharged from the basidia, forming the deposited spore mass. The sixth order, **Gasterales,** includes fungi with closed fruit bodies within which spores are formed; these, however, are not activity discharged from the basidia and fail to form any spore deposit.

The Gasterales do not form a developmentally homogeneous group – in fact it is an extremely heterogeneous grouping. Perhaps they represent the termination of several developmental lines of different origin. Different views are therefore supported in their classification, manifesting themselves in the unequal evaluation of the individual groups. Many mycologists regard the group discussed here as an order (Gasterales), others (including the author) as a higher category – a class (Gasteromycetes).

The order **Aphyllophorales** (non-gilled fungi) have either a resupinate fruit body spreading out loosely on the surface, or one of a definite shape. Resupinate species are most frequently found on the underside of uprooted tree trunks and fallen branches, on putrescent wood logs, in stump cavities, etc. So far a definitive classification of this large order has not been worked out; several families including a great number of genera are distinguished.

The following survey again presents families included in this book:

Stereaceae are lignicolous fungi having tough, resupinate to pileate fruit bodies usually surviving for several years, often densely arranged one above the other.

Thelephoraceae have brown-coloured, mostly angular, echinulate or verrucose spores. This holds for the genera *Tomentella* and *Thelephora,* as well as for the pileate fruit bodies of Hydnaceae – such as the genera *Hydnellum, Sarcodon* and *Phellodon.*

A characteristic of lignicolous fungi of the family Ganodermataceae are also brown-coloured spores, but these are truncate and ornamented. The fruit bodies of Ganodermataceae are tough to ligneous.

The family Coniophoraceae has relatively large, coloured, smooth spores – e.g. genera *Serpula* and *Coniophora.*

Clavariaceae and Sparassidaceae have erect, cylindrical or clavate, simple or ramose fruit bodies. The hymenium covers most of the outer surface of the fruit body. There are a large number of genera including *Clavaria, Ramaria* and *Sparassis.*

Hydnaceae and Auriscalpiaceae are fungi with an echinate (spiny) hymenophore – for example *Auriscalpium, Hericium, Hydnum.*

The family Poriaceae includes the greater part of polypores whose tubular hymenophore is inseparable from the tough context to which it is firmly accreted. Besides many true polypores (e.g. *Albatrellus, Fomes, Laetiporus, Trametes, Piptoporus*) the genus *Fistulina* also belongs here, though it is often classed with the independent family Fistulinaceae. A part of the polypores (the genus *Polyporus*) has been transferred as the family Polyporaceae into the order Polyporales, together with several genera bearing a lamellate hymenophore but similar in anatomical structure.

Cantharellaceae and Lentinellaceae are fungi with a distinctly developed fruit body whose underside is covered by the hymenium; this surface is either smooth or augmented by blunt, shallow, often anastomosing ridges, gill-folds or gills. From the well-known genera belonging to this group, which are collected for table use, let us mention the Chanterelle (*Cantharellus*), *Craterellus* and *Gomphus,* sometimes classed into the independent family Gomphaceae.

The hymenophore in the Schizophyllaceae is formed by pseudogills splitting in two along the edge.

The order **Polyporales** includes fungi having a lamellate or porous hymenophore. The fruit bodies are mostly annual, fleshy when alive, often tough in dry weather, resilient or almost hard as wood, and with a central, eccentric or lateral stipe. The hyphae are relatively thick-walled. The spore print is white, cream-coloured or lilac (greyish violet). They are mostly lignicolous fungi. The families Polyporaceae and Pleurotaceae belong to this group.

As a general rule, fungi of the order **Boletales** have a tubular hymenophore covering the lower surface of the cap as a continuous layer, quite easily separable from the cap context. The fruit bodies are fleshy in most cases, often with a developed covering (velum), usually with a central stipe. They grow on the ground, most frequently in mycorrhizal association with woody plants. The following three families are included in this book:

In the true Boletaceae the velum is either present or absent, the cap cuticle is either viscid or dry, smooth or matted-tomentose. The pores are white, yellow, olive-hued, pink or red, sometimes staining blue, green or red when bruised. The spores are ellipsoid to elongate-fusiform, mostly smooth. They are mycorrhizal fungi. Best known are the genera *Boletus, Tylopilus, Suillus, Xerocomus* and *Krombholzia.*

Paxillaceae are closely related to the foregoing family. Their hymenophore bears gills which are sometimes anastomosed or forked, relatively narrow, decurrent, yellow, orange or brown. The cap has an involute margin. The spore sprint is whitish, cream-coloured, yellow or brown; the spores are ellipsoid, ovoid, cyanophilous, nonamyloid. The genera *Paxillus, Hygrophoropsis* and *Omphalotus* belong here.

Gomphidiaceae have thick, decurrent, distant gills, fleshy fruit bodies with a dry to markedly glutinous cap, a black or olive-black spore print, and fusiform, smooth, cyanophilous, nonamyloid spores. Cystidia are developed. The genus *Gomphidius* is sometimes divided into two genera – *Chroogomphus* and *Leucogomphidius.*

The order Agaricales — gill fungi has juicy or tough fleshed fruit bodies. Usually they are differentiated into a cap and a stipe which may sometimes be reduced so that the fruit body is directly attached to the substrate.

The spore print varies in colour; the spores are colourless or coloured sometimes smooth or ornamented. When young the fruit body is sometimes encased in a veil.

The gill fungi are usually divided into ten families.

Now let us study about these families.

Hygrophoraceae

Fruit body is soft juicy, covered with slime. Gills are thick and distant basidia is narrow and long, spores are colourless.

Tricholomataceae

This family may or may not consist, velum. They have deeply decurrent gills. Spore print is white or yellowish. Sometimes it may be pale pink or violet, spores are often amyloid.

Entolomataceae

The fungi which belong to this family vary in their colour and size. Sometime they may be entirely stipeless. The spores are mostly angular, elongate. The basic genus is Entoloma.

Pluteaceae

These fungi are fleshy in nature and have central stipe, with thin broad, crowded, gills. The trama in these fungi are inverse. In some species, a volva is developed, at the stipe base. The spore print is pale pink. Spores are smooth, they usually grow on wood.

Amanitaceae

These are characterised by fleshy fruit bodies when young the fruit bodies are enveloped in a velum leaving patches on the cap cuticle, and an annulus on the stipe. A sheath is often developed at the stipe base. The gills are free

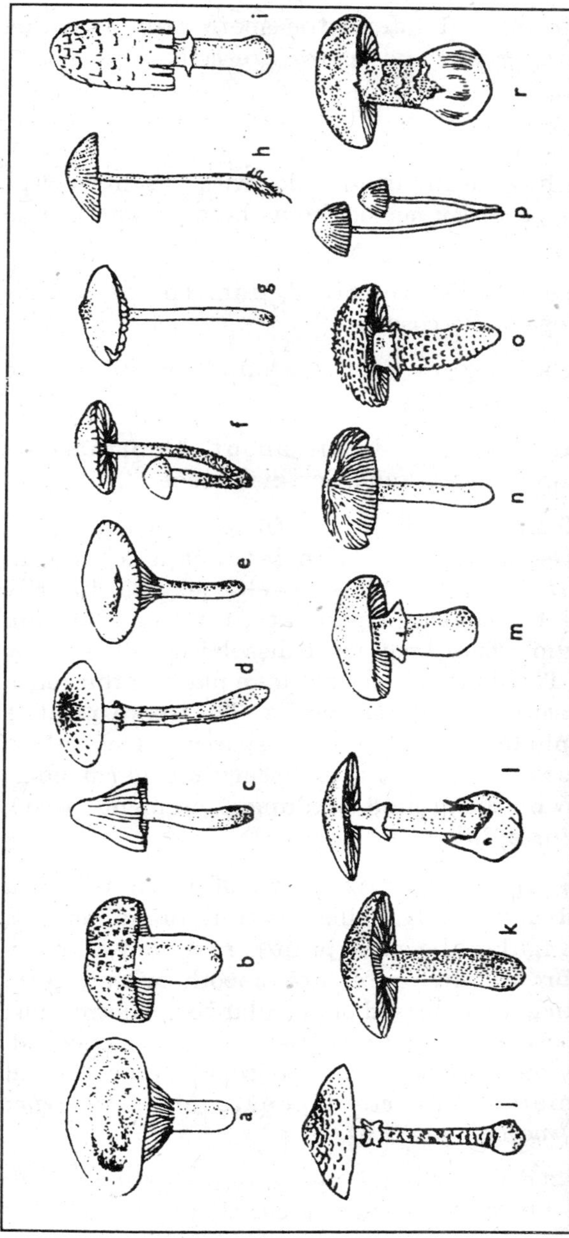

Examples of gill-fungi fruit bodies: (*a*) *Lactarius*; (*b*) *Russula*; (*l*) *Amanita*; (*m*) *Agaricus*; (*n*) *Inocybe*; (*o*) *Pholiota*; (*p*) *Psathyrella*; (*r*) *Cortinarius*. (*c*) *Hygrocybe*; (*d*) *Armillaria mellea*; (*e*) *Clitocybe*; (*f*) *Collybia*; (*g+h*) *Mycena*; (*i*) *Coprinus*; (*j*) *Lepiota*; (*k*) *Tricholoma*;

broad, the gill trama is bilateral the spore print is white, rarely linged with green. The spores are colourless mostly smooth.

Agaricaceae

These fungi have fleshy fruit bodies with a central stipe which can easily be disconnected from the cap context and bears an annulus.

The gills are crowded free broad, thin. The cap cuticle is often squamulose or floccose.

The fruit bodies grow on the ground, on dung, and humus.

Now let us study in detail about, Coprinaceae, Bolbitiaceae, Strophariaceae, and Cortinariaceae.

The fleshy fruit bodies of the Coprinaceae have a central stipe and thin, free or adnate gills; their trama is regular. In the genus *Coprinus,* fruit bodies develop extraordinarily quickly and, as the spores ripen, at first the gills and ultimately the whole fruit body start dissolving into a pulpy mass (autolysis). This process does not take place in the genea *Panaeolus, Panaeolina* and *Psathyrella.* The spore print is dark brown, purple-brown to black, the spores are smooth or ornamented, mostly possessing a discernible germ pore. Coprinaceae grow on the ground, on dung, excrements, wood, and on the remains of herbs or grasses.

Bolbitiaceae are smaller-sized and often thin-fleshed fungi with a central stipe. The gills are thin, broad, adnate, the gill trama is regular, the spore print is rusty brown, rusty ochre to umber-brown, the spores are smooth, with a germ pore. The cap cuticle is composed of globular to pyriform cells; cheilocystidia of a characteristic shape are often developed. Bolbitiaceae grow on excrements, on the ground, on wood and other plant remains. Important genera are *Conocybe, Bolbitius,* and *Agrocybe.*

Variously sized fleshy fungi with a central stipe belong to the family Strophariaceae. The cap cuticle consists of long hyphae, the sometimes distinctly developed velum leaves a ring or other remnants on the stipe. The spore print is pale violet or tinged with various shades of brown. The spores are smooth, mostly furnished with a germ pore. Cheilocystidia and also pleurocystidia are developed, some genera have chrysocystidia. Strophariaceae grow on the ground, often on decaying wood or on other plant remains, sometimes also on excrements. The main genera belonging here are *Stropharia, Hypholoma, Psilocybe* and *Pholiota.*

Cortinariaceae greatly resemble Tricholomataceae and represent the second largest family of gill fungi. The characteristic feature common to all Cortinariaceae is the colour of the spore print which, as a rule, is brown or rust-coloured; it is white only exceptionally. The spores are often ornamented, verrucose, and usually lack a distinct germ pore; the cap cuticle is composed of radially arranged hyphae; the gill trama is regular. The universal veil (*velum generale*) is either developed or absent; the partial veil is usually present in the form of a cortina. Cystidia are often present. Cortinariaceae grow on humus, wood and moss, and many species form ectotrophic mycorrhizae. The name of the family is derived from that of the most plentiful genus of gill-bearing fungi, *Cortinarius,* comprising approximately 500 species and divided into several subgenera which were once considered to be independent genera. Of the other large genera belonging to this family, let us mention *Inocybe, Hebeloma* and *Galerina.*

The order **Russulales.** Originally the Russulales were regarded as a family of the order Agaricales; at present, however, they are taken to be an independent order. A characteristic feature here is the structure of the fruit-body context containing – besides long, cylindrical hyphae – groups of globular cells (spherocysts) nidulariaceously arranged in its tissue. It is the spherocysts that cause the characteristic friability, that is the pomaceous brittleness of the broken context. Lactifers are also present in the context. In the *Lactarius* species they are filled with latex, i.e. a milky-white juice oozing from bruised places in fresh fruit bodies. The cap cuticle is often variegated; the spore print is white to bright yellow; spores are broad, subglobose, and broadly ovoid or ellipsoid, with a conspicuous strongly amyloid ornamentation on the surface. Cystidia are present in the hymenium. The order has a single family – the Russulaceae, including the genera *Russula* and *Lactarius* represented in the European forests by a large number of species. For the most part, they are mushrooms living in ectotrophic mycorrhiza with woody forest plants.

The class of **Gasteromycetes** (stomach fungi) includes fungi with an angiocarpous, predominantly globular fruit body which is either permanently closed or opens only in connection with the maturation of basidiospores. The fruit body consists of an external cover (peridium) and the part enclosed within (the gleba) containing the spore-bearing tissue. The gleba is often composed of cavities whose walls are partly formed by thick-walled hyphae; in the mature fruit body, these hyphae are transformed into the capilitium, while the other hyphae of the gleba disintegrate simultaneously with the basidia. The basidia are distributed throughout the fruit body either evenly or like birds' nests; sometimes they line the inner walls of the chambers or cavities in palisades. The development of the gleba is typical of individual families and genera of the Gasteromycetes. Ripe spores are not actively discharged (as for instance in the gill fungi).

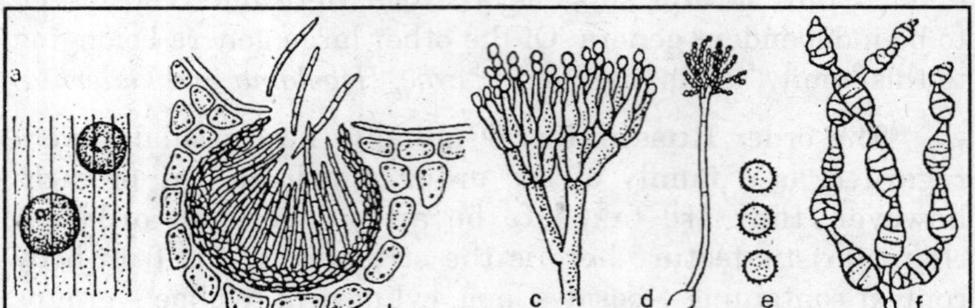

Example of some microscopic fungi of the group Deuteromycetes: *(a) Pycnidia of Phoma* sp.; *(b) Section through the pycnidium of Septoria* sp. *with conidia; (c) conidiophore of Penicillium* sp. *with conidia; (d) Conidiophore of Aspergillus niger with conidia; (e) Conidia of Aspergillus* sp.; *(f) Conidiophore of Alternaria tenuis with conidial chains.*

Phallaceae are conspicuous and sometimes very ornamental fungi. The young fruit body is egg-shaped. The peridium is often composed of three layers, the middle layer being thick and gelatinous. At maturity it is pierced through at the apex by the receptaculum (i.e. the infertile part of the gleba). The receptaculum carries the fertile part upward, while the rest of the peridium remains at the base in the form of a cup (volva). At maturity the fertile part of the gleba covering the surface of the receptaculum decomposes into a pulpy matter emitting an unpleasant odour. This odour attracts insects and these subsequently disperse the basidiospores contained in this matter. A number of tropical genera belong here, the genera indigenous to Europe are *Phallus* and *Mutinus*. The genera *Clathrus, Anthurus, Lysurus* and *Dictyophora* are sometimes classed with the independent family Clathraceae.

Hysterangiaceae have hypogeous, tuberiform or sub-globular fruit bodies often overgrown with mycelial threads, and a cartilaginous or gelatinous gleba composed of a tissue radially developing from the infertile base or from a gelatinous columella. The spores are smooth, ellipsoid or cylindrical, and are almost colourless. The genus *Hysterangium* belongs to this family.

Earth stars (Geastraceae) are terrestrial, and in the tropics even lignicolous fungi. When young, the globular or pyriform fruit bodies are usually embedded in the ground. In mature fruit bodies the outer peridium bursts and lifts the fruit bodies above the ground. The outer peridium splits into rays, in the centre of which the inner peridium with a globular gleba is seated. The gleba contains the capillitium and spores. The Geastraceae often remain in their spot for a long time.

Lycoperdaceae are mostly epigeous (at least at maturity). They are globular, pyriform or clavate in shape; the gleba fills the whole interior of the fruit body or passes over into a sterile tissue in its basal portion. The peridium consists of a thin and readily decomposing outer layer (exoperidium), and a usually thin, papery layer (endoperidium) opening by a small peristome or irregularly decomposing. In the members of the genus *Geastrum* it splits to form regular rays, the endoperidium is either attached to the exoperidium or is borne on a pedicel; the peristome is situated at the apex of the endoperidium. The capillitium is usually developed. Spores are globular, small, and coloured. The main genera include *Lycoperdon, Bovista, Calvatia, Langermannia,* and *Geastrum. Langermannia gigantea,* whose fruit bodies grow up to 50 cm in diameter, are among the largest of fungi.

Sclerodermataceae have tuberiform, fleshy, sometimes stipitate fruit bodies developing on or under the surface of the ground. The peridium either opens by an apical pore or bursts irregularly. At maturity the gleba is reduced to powder – either directly, or following an intermediate disintegration into pip-like structures. The basidia are either regularly distributed in the gleba, or form nidi, or fill in the cavities. Spores are globular, ornamented, and coloured. The best-known genera of this family are *Scleroderma* and *Pisolithus.*

Closely related to the Sclerodermataceae are Astreaceae, even though, in appearance, they markedly resemble members of the genus *Geastrum* to which they are not linked.

Nidulariaceae have globular or cup-like, sessile fruit bodies with a chambered gleba disintegrating into single peridioles. These either remain attached to the inner wall of the peridium, or are discharged. The basidia form the hymenium covering the walls of the peridioles, or assemble into nest-like structures situated within the gleba. Spores are ellipsoid, smooth, and colourless. The Nidulariaceae thrive on decaying plant remains or on old dung. On the whole, they are small fungi interesting both in appearance and in biology.

Deuteromycetes

This vast group includes all fungi characterized by purely asexual reproduction (without any foregoing nuclear association either by the decomposition of hyphae, or by the formation of special spores, known as conidia). They are known only in this form (nowadays termed anamorph), or demonstrably belong to other fungus groups – mostly to Ascomycetes – as their particular developmental stage. They mostly involve minute organisms which, for the most part, can be distinguished only under the lens, and only a first-rate microscope makes possible their identification and study. They occur practically everywhere, as saprophytes or parasites. Approximately 20,000 species have ben described hitherto. They are divided into two main groups – Coelomycetes and Hyphomycetes. In the Coelomycetes, conidia are borne on conidiophores or arise directly from conidiogenous cells growing out of the sterile hyphal tissue which forms closed on open fruit bodies. Several hundred genera including many thousand species belong to this group.

In the Hyphomycetes, conidia arise either directly from the mycelium or, most frequently, they originate on conidiophores from specific, so-called conidiogenous cells. Conidiophores grow freely on the substrate as fibrillose, hirsute or matted-tomentose layers. These so-called pseudomoulds are of considerable importance in phytopathology, technical mycology, veterinary science and medicine as organisms causing various diseases in plants, animals and man. The colouring of conidiophores substantiates their division into colourless ones (Moniliales or Moniliaceae) and dark-coloured ones (Dematiales or Dematiaceae). Of the most important genera at least the following should be referred to: *Aspergillus, Penicillium, Botrytis, Monilia, Cladosporium* and *Fusarium*. Also the Hyphomycetes include a great number of genera and species.

7

Edible Mushrooms

It is very difficult to say, which mushrooms are edible and which are not. The findings obtained in the course of the last decades reveal the necessity to be reserved and evaluate individual fungus species, on the bases of much stricter criteria.

The fungus species is considered edible if eating it causes no health disorders, provided it has, been properly cooked.

It is recommended to refrain from eating any raw fungi whatsoever. The ingestion of even small portions of raw fruit bodies of certain current mushroom species which may be eaten quite safely after cooking can cause grave nausea.

For example Honey fungus causes grave nausea.

The time needed for cooking varies from species to species, since it depends on the consistency of the flesh.

In fruit bodies with soft flesh it takes about 10 minutes, for together flesh at least 20 minutes are necessary.

Only faultless young or recently developed fruit bodies should be gathered for the table; they must be neither overripe and flaccid, nor infested with, microorganisms, mould or insect larvae.

Fruit bodies covered with a slimy and viscid cuticle, such as in usual e.g. in the *Suillus* species, must be cleaned and deprived of their pellicle in the very place where they have been found: this saves us a lot of extra work when cleaning them at home. It is also advisable to avoid collecting flabby, large caps of some *Leccinum* species unable to withstand transport. The collected mushrooms are put into a wicker basket. Plastic bags are absolutely inappropriate for the purpose, as the fruit bodies deposited therein get crushed and overheated. At home it is best to keep the mushrooms in a refrigerator at a temperature of 5-8ºC; here they remain in a good condition for 24 hours, but desiccate more rapidly.

The dish prepared with mushrooms should be eaten at once if possible, its further preservation (in the refrigerator of course) and reheating reduces its goodness and substantially affects its edibility. Mushrooms should be prepared for the table according to well-tried recipes, or it is possible to enrich the diet by using some species to prepare special dishes accentuating their taste qualities. For example, Saffron Milk Cap (*Lactarius deliciosus*) is best when pickled in vinegar, prepared as goulash or roasted on butter; puffballs are appetizing in soups; caps of *Lactarius volemus* and of the Parasol Mushroom taste excellently when roasted on fat, etc.

If we refuse to risk poisoning and the often ensuring serious damage to health, we must observe the basic rule – namely, that only the species we are quite reliably aquainted with should be collected. If, in spite of everything we are keen on enlarging our assortment and knowledge by further species, it is best to contact some specialist who will identify the mushrooms for us or confirm the correctness of our identification. It is also expedient to visit mycological advice bureaus functioning in certain cities, or else it is possible to attend special lectures and courses organized by mycological associations or clubs. The identification based exclusively on comparison with coloured plates is often rather unreliable (in view of the great colour and shape variability characterizing many species, of a possibly inaccurate reproduction of colours, etc.). It is always only a detailed, exhaustive description that can be conclusive and decisive.

The digestibility of mushrooms is directly related to the cellular structure of their bodies. Mushrooms are less digestible than other plants due to the fact that their cell walls are largely composed of the polysaccharide chitin resistant to the action of digestive fluids in the human organisms. The cellular plasma contains 70-95 per cent of water and only a negligible amount of sugars and fats which represent the substantial proportion of nutritional substances. The mushroom flesh contains about 3-9 per cent of proteins, their compounds being amino acids and peptides. Altogether, about 20 various amino acids were found in mushrooms; their content varies even in individual specimens of the same species. Sugars represent 1-6 per cent of the content, for the most part in the form of polymers, but there is also glycogen, mannitol, traces of sorbitol, arabitol, and others. Mushrooms further include 0.5-3.5 per cent of fats, mostly glycerides and glycolipides, less frequently phospholipides (e.g. lecithin), often in the form of oil droplets within spores or in the tissue. Also vitamins are present in mushrooms, especially the provitamin carotene, vitamins of the vitamin-B group, vitamin D, E and K in small quantities, and traces of vitamin C. Mushrooms also contain various enzymes and substances with antibiotic effects. The mineral substances contained in mushrooms are first of all potassium (K), phosphorus (P), calcium (C) and iron (Fe). The capacity of fungi to accumulate from the soil these mineral substances (trace elements), which are so important for the enzymatic processes taking place in the human organism, presumably exceeds that of green plants. The concentration of trace elements in fungus fruit bodies depends on the habitat.

Recommendations as to which edible mushrooms should be collected are always a problematic matter, connected with various circumstances among which tradition, conservatism, individual taste, etc. play a major role. In general it can be said that in countries where wild edible mushrooms are not habitually collected, preference is given to artificially produced field agarics. Among the mushrooms gathered in their natural state, high value is attached to those belonging to the Boletaceae family: in most cases they are easily recognized and endowed with a pleasant smell, taste and consistency. The most popular and generally known bolete is of course the Edible Boletus or Cèpe (*Boletus edulis*). Otherwise it is recommended to eat mushrooms in mixtures in which the specific taste qualities, consistency and smell of the single species come to the fore. Species with a less distinct taste are added to strongly aromatic or more acridly tasting species. An adequate preparation in the kitchen renders the mushrooms more palatable and more digestible. Tougher species are cut into thin slices, dried mushrooms ground into powder can be used as spices, many species can be preserved in salt or pickled in various solutions. The trimmings added to mushroom dishes should be easily digestible – e.g. potatoes and vegetable salads. Special literature brings a great number of recipes for everybody to choose according to his individual taste.

8

The Morphology of Edible Mushrooms

So far we have studied about some edible mushrooms, but as we learnt earlier which edible mushrooms should be collected are always a problematic matter.

Now let us study about some of the mushrooms individually which are edible, and also let us go through about their morphology and characteristic features.

Russula Integra

Russula integra is one of the most common russulas, found in early autumn, even if the weather is dry.

The fruit bodies are medium to large in size. The cap is 5-12 cm. in diameter, semi globular at first expanded convex later, slightly depressed at the centre, tough fleshed when young.

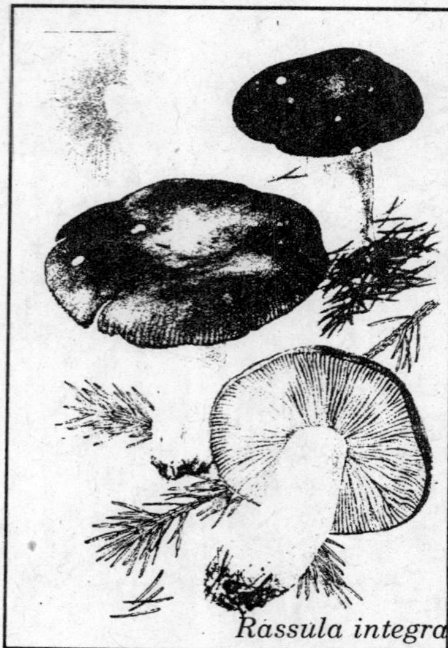

Rassula integra

Its colour may be dingy dark violet, redish brown, dull or violet red; sometime it may have chocolate or light brown colour.

The cap cuticle is glabrous remarkably shiny when dry detachable to one half of its radius.

The stipe is up to 7 cm. long and 2-3 cm. thick firm, white all along its length.

The gills are thick and broad. The flesh is white, unchanging with a mild taste, and has virtually no odour.

This species is extraordinarily variable in colour.

Russula integra is a tasty edible mushroom.

Russula Paludosa

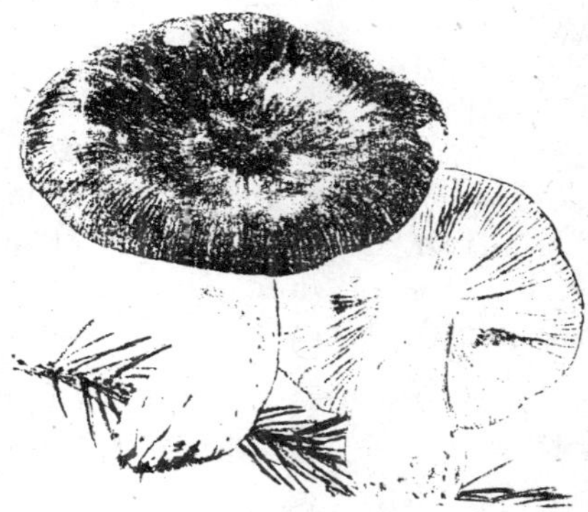

A large russula of a lofty build and pleasant red colouring – this is perhaps the most distinctive characteristic of this species. The 6-15 cm. wide cap is semiglobular when young, flatly expanded in age, beautifully red-coloured like ripe strawberries or rosehips. This colour often changes to orange or cream-yellow. The cap cuticle is glabrous, moist, shiny when dry, and readily peeling far towards the centre of the cap. The long (6-15 cm.) and thick (2-4 cm.) cylindrical stipe is white, very often tinged red or rosy on one side. The gills are rather crowded at first, becoming fairly distant later and alternating with numerous lamellules. They are up to 12 mm. broad, pale cream-coloured when young, then pale ochre, sometimes with reddish edges (especially near the cap margin). The flesh is relatively firm, permanently white, and has a mild, sometimes penetrating taste.

R. paludosa is a typical species of damp coniferous forests on acid soils, occurring mostly in pine woods and less often in spruce woods. It can often be found growing in moss or sphagnum, and in billberry, red-whortleberry and molinia undergrowths. Under favourable conditions, it appears in large quanities in later summer in foothill and mountain forests. It is among the most tasty of edible russulas and is very popular.

Russula Mustelina

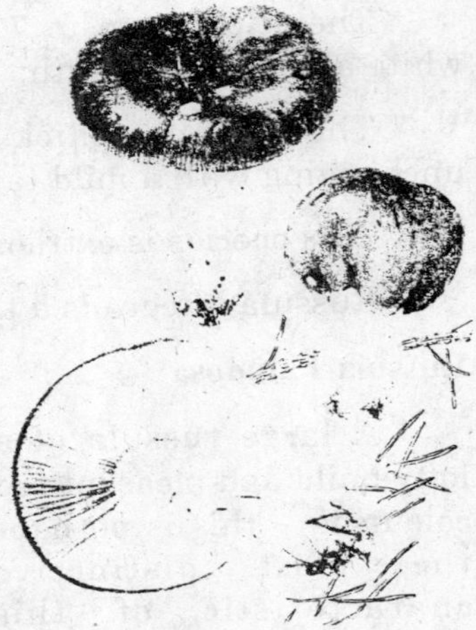

Russula mustelina is a robust firm fleshed species with a hard stipe and mostly brown coloured cap which may grow as much as 15 cm. in diameter.

The cap cuticle is smoky or yellowish brown slightly viscid and shiny when moist, matt when dry.

The stipe is 3-11 cm. long, 1.5-4 cm. thick, cylindrical, often crooked, white later slightly yellowish, gradually turning brown or rusty and becoming brown all over when old. The gills are thick crowded and thick.

The flesh is tough white readily turning yellow brown with age. It is almost odourless and has a mild to sweet taste.

Russula mustelina is confined to old coniferous forests.

Young fruit bodies show strong resemblance to the Edible Boletus by their semiglobular-convex cap and by its colouring.

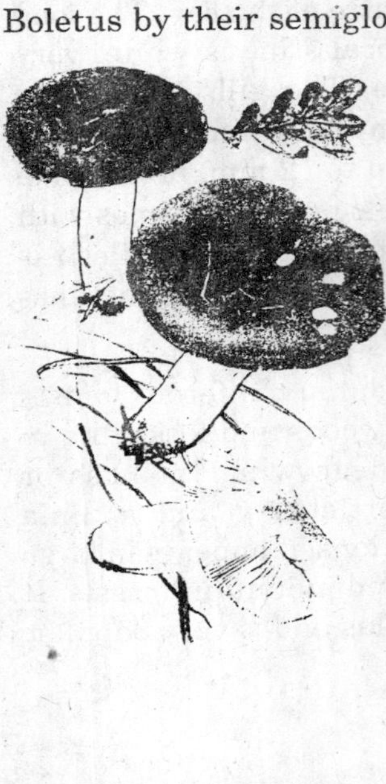

Russula mustelina is an edible mushroom of good quality. Sometime the cap of the stinking Russula has a deeply grooved margin and is predominantly ochre-yellow in colour.

Russula Xerampelina

Russula xerampelina was once considered to be an independent species principally on the basis of three features. The first is a marked trimethylamine smell (i.e. a smell or pickled herrings or boiled crayfish; another fungus with a similar smell is *Lactarius volemus*). The second distinguishing feature is the tendency of fruit bodies to turn brown on injury, the third is a typical chemical reaction with a 10 per cent green-vitriol solution. When touched with this, the flesh turns green. The is also why the Czech mycologists Melzer and Zvára, who elaborated a modern conception of the genus *Russula* in 1927, treated this species as belonging to the independent section *Viridantinae.*

R. xerampelina is a strongly variable species, and there seems to exist no colour hue absent from its cap cuticle. Furthermore, mycorrhizal relationship have arisen between individual varieties differing in colour and the woody plants upon which they depend. This is why some present-day mycologists regard these deviations as microspecies. The well-known French mycologist H. Romagnesi, a specialist in the genus *Russula,* distinguishes 15 independent species belonging to the section *Viridantinae.*

In the prevailing opinion, the typical form of *R. xerampelina* is a relatively robust russula with a bright reddish purple or carmine cap, sometimes tinged with black in the centre, about 10 cm in width, with a mostly pinkish red stipe, occurring in coniferous forests. This species, like all the others included in this group, has a mild, nonacrid context and is edible.

Grass Green Russula

It is difficult to differentiate between these fungi even after a careful comparison of exact descriptions of both genera in question Russula and Amanita.

The Grass Green Russula has a cap up to 10 cm. wide olive green, sometimes lighter with a vague uchre, brownish or whitish patches so that the green colour is sometimes more or less suppressed.

The cap margin is relatively thin sharp edged and striate.

The glabrous cuticle tacky when moist is readily separable almost to the centre.

The stipe is 5-8 cm. long, 1-2 cm. thick, firm fleshed, cylindrical, often tapered at the base, it is pure white but tends to turn yellow or brown particularly at its basal portion. The gills are thin and crowded.

The originally white flesh becomes rusty on injury, or with age.

The Grass Green Russula occurs abundantly not only in the spruce and pine forests, but also in grass along woodland tracks and in parks, predominantly in acid soils. It grows throughout the summer and is ediable.

Russula Olivacea

This is the most robust russula inhabiting European forests. Its cap grows up to 20 cm. in diameter; it is shallowly convex, flat to broadly depressed later, with a dull, almost velvety cuticle separable at the margin only. It is very variable in colour. In some specimens purplish red or violet predominates, in others olive-green tinged with ochre or brown. There is, however, one feature characteristic of this russula; the pigment in the cuticle along the periphery of the cap is distributed in flexuous narrow bands. The 6-10 cm. long and 2-5 cm. thick stipe is robust, cylindrical, firm at first, later spongy, tinged with purple either all over its surface or at least on one side, slowly turning yellow to brown in the basal part when bruised. The gills are remarkably broad

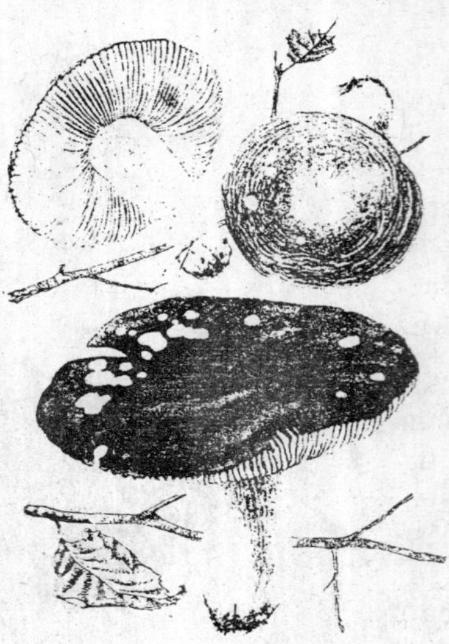

(up to 1.5 cm.), distant, thick, brittle, soon completely or partly bright egg-yolk-yellow with red-coloured edges. The flesh is white, rather dry, inodorous, and tastes bland. When touched with a drop of phenolic (carbolic) water it instantly produces a reddish violet stain.

Russula olivacea occurs in the warmest summer months especially in rather dry, sunny forests or in their outskirts. It grows under spruces and pines, but also in beech forests, usually on relatively acid soils. It is edible and tasty although its flesh is much too dry.

Bare-toothed Russula

The Bare-toothed Russula is one of the most savoury mushrooms. The whole fruit body has a relatively firm flesh. The cap is 6-11 cm. wide covered with a dingy reddish fleshy cuticle weathering to grey or bleached in places.

The stipe 3-10 cm. in length and 1.5-3 cm. in width, and is often much shorter than the diameter of the cap. It narrows downwards, with rusty spots on white surface.

The cortex of a fresh fruit body soon develops a taste of hazel nuts but is almost inodourous.

The spore print is pure white. The reaction of the Bare-toothed Russula to the green-vitriol solution is extraordinarly intense; one drop of the solution instantly colours the stipe surface or the flesh to orange red or pinkish orange.

The Bare-toothed Russula occurs very abundantly from the end of May throughout the summer in forests of all types. In some regions it is most common in oak forests on non-calcareous substrates.

Russula Cyanoxantha

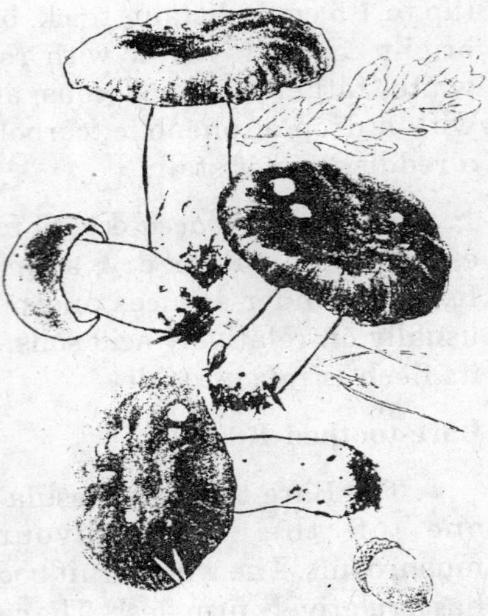

Some *Russula* species show extraordinary variability in the colour of fruit bodies. This applies also to *Russula cyanoxantha*. In some of its forms, the cap cuticle is blurred cloudy violet or bluish purple. Very often, however, specimens are found in which green predominates – the cap centre especially is often dark green; in still other cases the cap may be partly pinkish beige to lilac. But despite these fairly large colour variations, this species can safely be distinguished by one characteristic feature; if we let our fingers pass over the gills, these seem greasy, pliable, and do not break as in other russulas. The relatively short and thick stipe is usually pure white all over (more rarely rosy with a violet tinge), and also the gills are persistently white. If the cap of a fresh fruit body is cut off from the stipe and placed on a sheet of black paper with the gills facing downwards, spores resembling chalky white powder may be found to have dropped on the sheet in the course of several hours. The colour of the spore print is a significant distinguishing feature on the basis of which individual species of *Russula* can be identified. Another essential feature for distinguishing this species is the finely wrinked cap cuticle and the pleasant, non-acrid taste of the odourless flesh which shows no colour reactions in a 10 per cent green-vitriol solution.

R. cyanoxantha is a good edible mushroom. It appears as early as the end of spring, particularly in deciduous forests but later also in coniferous forests. It is distributed throughout the north temperate zone and in north Africa.

Common Yellow Russula

In late summer and early autumn, the monotonous spruce forest floor covered only by needles or a sparse

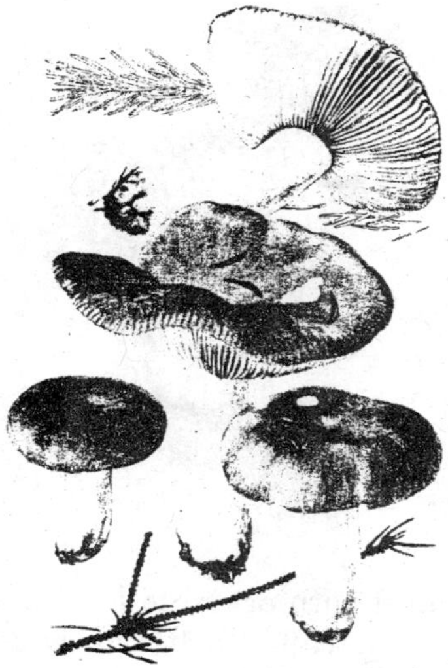

undergrowth is enlivened with yellow fruit bodies of russulas.

Its cap is coated with a yellow sometimes almost golden yellow cuticle, rather shiny when dry, sometimes slightly wrinkled or slightly venose, readily peeling almost to the cap centre.

The white or more rarely yellowish stipe is 3.5-7 cm. long and 1-2.5 cm. thick, tough at first spongy elastic later, occasionally with a clavetly expanded base, often covered with minute scales or brownish granules.

The gills white at first become cream-coloured with age sometimes with a lemon-ochre tinge staining light brown on handling especially at the edge.

The flesh has an inconspicuous faintly acid fruity smell and a rather acrid-rarely mild-taste. The spore print is white or cream.

The Common Yellow Russula occurs not only in coniferous forests but also in deciduous forests.

Inspite of its acrid taste this species is collected by some mushroom pickers to flavour bland mushroom mixtures.

Cracked Green Russula

The green colour in gill fungi is invariably regarded as a warning signal by all mushroom pickers who, though not safely acquainted with the individual species and hence incapable of reliable identification, are keen on gathering mushrooms for culinary purposes. This is because the most poisonous fungus, the Death Cap, has a green-coloured cap. Many a mistake with fatal consequences has been caused by

the failure to distinguish the Cracked Green Russula or some other similarly coloured species (e.g. the Grass Green Russula) from the Death Cap. However, a more experienced mushroom picker who can make a clean-cut distinction between *Russula* and *Amanita* is in a position to enrich his collection by green *Russula* species as well.

One of these is the Cracked Green Russula, readily distinguished by its verdigris or greyish green cap, fading to pale buff with age, and having a characteristically areolate cuticle. The cap is 5-12 cm. wide; the stipe is 3-9 cm. high, 2-5 cm. thick, cylindrical. The spore print is white or whitish. The whole mushroom is relatively stout and firm-fleshed; its cap tends to remain semiglobular-convex for a long time. The gills, stipe and flesh are white but susceptible to staining rusty. The flesh of a fresh fruit body, when touched with a drop of green-vitriol solution, immediately turns deep red.

This edible and, moreover, delicious russula is abundant in summer and autumn in relatively dry and sparse deciduous and coniferous forests.

The species is also used in industry: it is dried on a large scale for the purposes of mushroom-extract production. It has also proved suitable for flavouring soups and sauces.

Wood Witch, Stinking Polecat, Stinkhorn

This conspicuous and well-known fungus has become wide spread, and in some regions it is almost becoming a weed fungus.

Wood Witch is flooding the forests in vast multitudes and attracting attention from afar by the typical vile odour of its ripening fruit bodies.

It occurs in woods of all types, but also in gardens and parks. Young stages, the well known elastic eggs similar to tennis balls are about 6 cm. in size with characteristic tough mycelial rhizomorph at the base.

Their faint odouris suggestive of raddish. The cap which is covered with an olive green slime emitting a nauseous smell, which attracts numerous insects, particularly flies, that subsequently help to disseminate its spores.

The range of the Wood Witch covers the entire temperate zone of the old world.

The young still tough and closed eggs of the Wood Witch are good when fried although most people prefers to avoid, this fungus or unnecessarily destroy it.

Hysterangium Graveolens

Hysterangium graveolens is one of the most common representatives of hypogeous gasteromycetes. There exists a large number of such fungi, and their correct identification is a distinct problem for the specialists. This applies in particular to the genera *Hymenogaster, Rhizopogon, Melanogaster* and others, whose discrimination meets with difficulties because the fungi involved are given to extraordinary variation in both macro- and microcharacteristics.

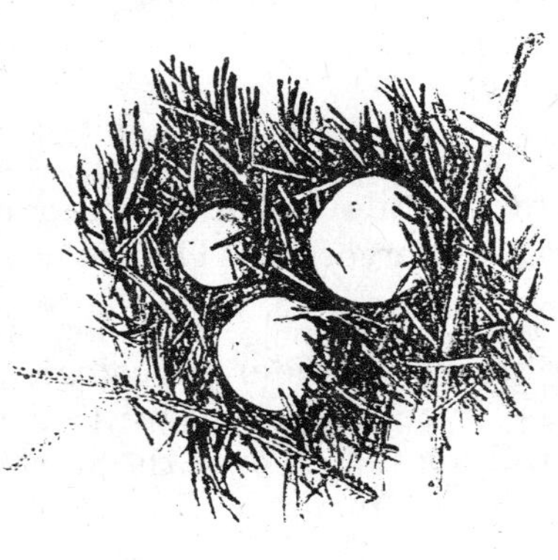

In general, fungi of the genus *Hysterangium* can easily be disginguished by their readily separating peridium. From the tuberous fruit bodies – brown with a whitish, greyish and later reddish tinge, 1-3 cm. in size – the peridium sloughs off in large, thick patches, revealing an olive-green, relatively tough and elastic gleba. A section through the gleba clearly reveals both greyish, jellified and dendriform veins branching off from the common central column or columella, and narrow and flexuous veins. Fruit bodies have a white mycelial rhizomorph at the base and grow, for the most part, caespitosely or gregariously right under the surface of the ground, most frequently in spruce forests. At maturity they partially emerge above the surface.

H. graveolens is commonly found along woodland paths. It is an edible fungus but, due to its very small size and relatively rare occurrence, it is of no great, importance for mushroom pickers. It is often introduced in literature under the name *H. clathroides* or *H. separabile.*

Giant Puffball

A Giant Puffball is always a notable find especially where a fruit body of a maximal size and weight is concerned.

These fungi are found either in human settlements or in their vicinity; in gardens, lawns, meadows, on slopes, and in pastures.

The exoperidium of the Giant Puffball gradually turns yellow and peels of in pieces while the gleba is tansformed into yellowish brown cottony lumps.

Being undifferentiated into a fertile and a sterile part the whole interior of the fruit body undergoes this change once the spores begin maturing.

The Giant Puffball also known in older mycological literature under the names Lycoperdon giganteum, *L. maximum* and *L. bovista* is widespread all over the world; it grows even in the tropics and in the far north.

At an early stage while the fruit body is still white both on the outside and inside.

It is edible and is probably the most palatable of all puffballs.

Common Puffball

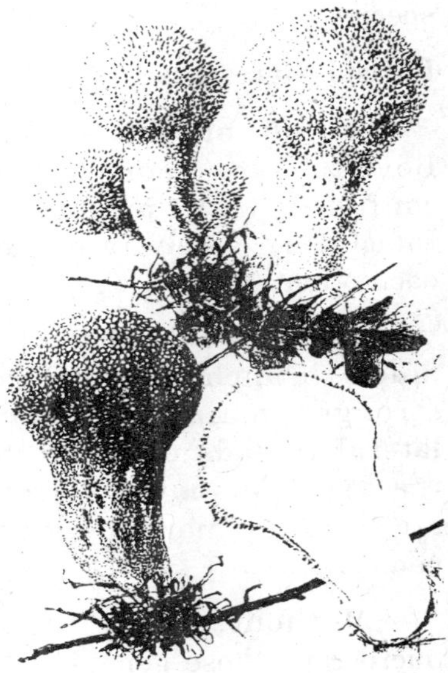

This is one of the most abundant puffballs commonly found in forests of all kinds. Its fruit bodies grow up to 8 cm. in height and 5 cm. in width; they are pear-shaped or club-shaped, often growing gregariously on the ground among fallen leaves and needles as well as in grass. The young fruit body is entirely white, densely coated with spines and readily detached conical warts leaving characteristic angular patches (areoles). With the ripening of spores, the soft, porous white flesh of the gleba gradually becomes yellowish at first, then greyish green and finally brown, turning into a brown spore powder. This is released through a small pore situated at the apex of the fruit body. The capillitial threads are only sparingly dichotomically branched, 4 mm. thick, thin-walled, yellow-brown. The fresh context of the gleba has a pleasant mushroom-like smell and a mild taste.

The collection of quite young, still white fruit bodies of this puffball for table use can be fully recommended: they are suitable for all culinary purposes, particularly for flavouring soups.

Lycoperdon pyriforme, visibly obviously different from the above by the minutely verrucose or, rather, farinaceous surface coating, grows on decaying wood, especially on stumps where its fruit bodies are often aggregated in dense tufts. It grows out of clearly visible white mycelial threads penetrating deep into the rotting wood. Young fruit bodies are also edible.

L. echinatum, which grows in summer and in autumn mainly in beech or mixed submontane forests, is also an edible species.

Bovista Nigrescens

Fungi of the genus Bovista are sometimes confused with puffballs. Infact they closely resemble each other.

In Bovista hyphae are much branched, with a stronger main axis and lateral threads tapering to the apex. Moreover spores are borne on markedly long pedicels.

Without using the microscope these genera can be distinguished by the presence or absence of the sterile base in the fruit body.

Like the other species, Bovista nigrescens grows outside the forest, in meadows and pastures, on grassy slopes, along field paths, and also on lawns of urban parks.

It is a robust fungus growing upto 5 cm. in size; its nature endoperidium ranges in colour from blackish brown to almost black.

Closely related is *B. plumbea,* occurring predominantly in lowlands. It is smaller, measuring only 1-3 cm. in diameter, its endoperidium is pale grey or bluish grey.

Young fruit bodies still white both on the outside and inside, can be collected for eating.

Morchella Elata

It is one of the largest species of the genus, the fruit bodies being 10-25 cm. high and as much as 8 cm. wide. It differs from the Common Morel not only in size but also in its elongate-cylindrical, but sometimes only slightly pointed, globular cap whose colour is a shade of brown or more rarely pink to reddish purple. The main ribs of the cap are rather thin and more or less parallel, connected by thin, narrow, transverse and oblique ridges; elongate, tetragonal pits are thus formed between the ribs which are closed at the cap margin. The longitudinally pitted or grooved stipe is usually strongly inflated into a bulbous base as much as 8 cm. thick.

Morchella elata grows in April and May, and on higher ground also in June. It is sporadically scattered on the ground of deciduous and coniferous forests, in gardens, in old parks, and very often on rubbish heaps in urban yards. It is edible.

Another large species, the Prague Morel (*M. pragensis*), occurs in places greatly affected by human activity, e.g. in factory yards, building sites, dumps, railway side-tracks, and often in the very centre of large industrial cities. The Prague Morel has a bilaterally compressed cap resembling a ping-pong bat, considerably elongated, and yellow to olive-brown. The size of its fruit bodies varies within the range of 6-28 cm. in height and 4-8 cm. in width. It is also edible.

Pisolithus Arrhizus

This fungus is related to the genus scleroderma. The peridioles of which it is composed are located in a shiny black

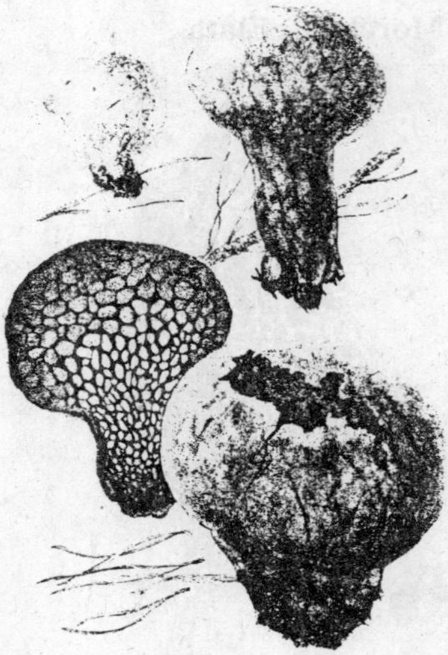

case; they are yellow at first and brown when older and contain basidia with spores.

At maturity the whole interior of the fruit body disintegrates into a strongly pulverous, chacolate-brown matter formed by the spore mass from the crumbled peridioles.

This is released through the upper part of the peridium which has fallen to pieces.

The sterile, thick and branched basal part of the peridium resembles a root. It is sometimes as much as 30 cm. long embedded deep in the ground where it remains long after the disintegration of the fruit body.

This fungus is most commonly found on sandy soil in pine forests. Here it sometimes grows along tracks or on deep inclines of hollow ways torn down and grooved by rainfalls.

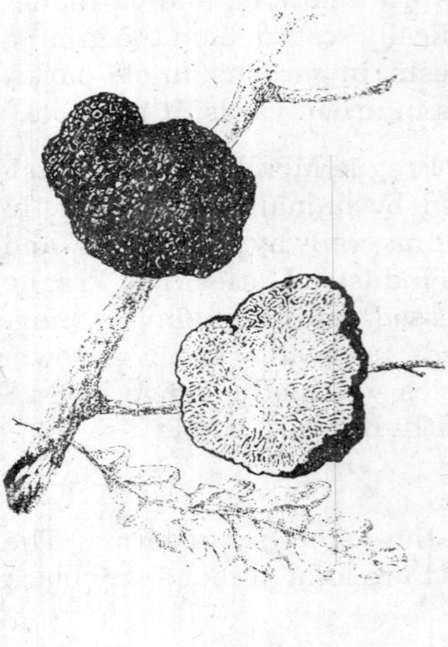

Young fruit bodies, are edible in small quantities.

Winter Truffle

This is one of the 'true' truffles that are highly esteemed from the culinary point of view. Although its aroma is not so intense as in the renowned Périgord Truffle (*T. melanosporum*), it exceeds the Summer Truffle in both aroma and taste. The tuberous fruit bodies of the Winter Truffle, ranging in size from a hazel nut to a human fist, are covered with low and broad, tetragonal to hexagonal, dark warts. The gleba is ashy grey, intertwined with dirty white veins. One to six ripe spores develop in shortly pedicellate, globose to ovoid asci and are densely covered with long spines.

The Winter Truffle is a subterrestrial fungus. It is confined to forest humus and lime-rich soils often overlaid with decaying foliage, particularly in warm oak forests; in some regions, however, it also appears in beech and hornbeam forests. It attains maturity in late autumn or in winter. In the process of ripening, the fragrance of its fruit bodies, pleasant at first, is gradually replaced by a nauseating odour, especially after keeping ripe fruit bodies in an enclosed space for several days. The Winter Truffle, abundantly distributed in some parts of France and Italy, rarely occurs in more northern regions of western Europe.

Common Morel

Morels are the most advanced Discomycetes and at the same time the best known cup fungi to be collected for their excellent taste.

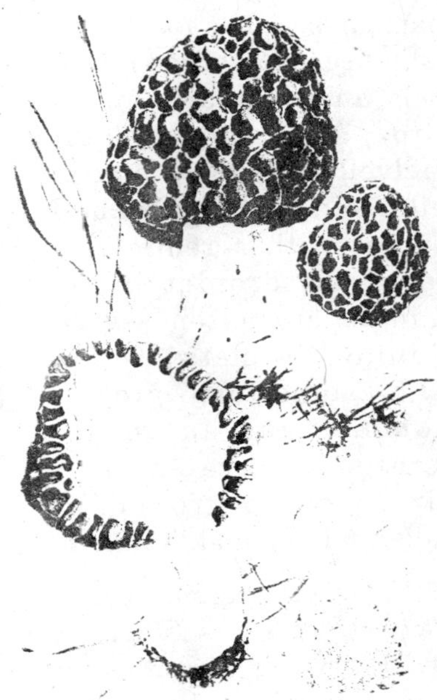

Morels are extremely variable in the shape, size and colouring of fruit bodies, and mycologists have not, yet arrived at a uniform evaluation of all these variations.

The cap of the common morel is attached to the stipe by its lower margin and is marked with longitudinal ridges of irregular shape.

The ridges are usually sinuous and are separated by deep intermediate pits.

These lacunae are rounded to rounded-polygonal in shape.

In its typical form the common morel has an avoid-ellipsoid to globular, sometimes conical.

It has even cylindrical cap, usually 4-8 cm. high and 4 cm. wide, brownish or greyish in colour. It grows in April and May in sunny places in forests, but also in lawns, parks, gardens, etc.

The fruit bodies of all morels are hollow and fragile. It is edible.

Common White Helvella, False Morel

Fungi of the family Halvellaceae belong to the highly developed Discomycetes. Their fruit body is differentiated into a sterile part (in the form of a stem or stipe) and a fertile part, bearing the sporiferous layer made up of asci and paraphyses. The fertile part is saucer-shaped or resembels an irregularly formed cap or saddle. All Helvellaceae have broadly ellipsoid, usually smooth spores with a remarkably large oil droplet in the middle and often several minute droplets around. The saddle fungi grow in summer and autumn in forests, on bare and relatively damp ground, in fallen foliage and in humus.

One of the most abundant species is *Helvella crispa,* which is commonly encountered in both deciduous and coniferous forests. It is easily recognized by its almost white or creamy yellowish fruit body, irregularly lobate cap and deeply grooved stipe. The outer cap surface is finely felted, in contrast to the very similar *H. lacunosa* whose cap has an entirely smooth outer surface. *H. lacunosa* is usually coloured grey, or greyish brown to blackish; there also exist almost white forms, but these are very rare. This species also grows on the ground in shady and rather damp parts of both coniferous and deciduous forests, and under shrub thickets. Both species are edible.

Early Morel

The Early Morel is a spring fungus. It can be easily identified by its cylindrical, hollow slender stipe with a campanulate cap sealed at its apex.

The cap attached to it by quite a small area at the very top. The cap surface is uneven, and undulate with rounded, longitudinal, yellow brown thick folds.

The fruit body reaches up to 15 cm. in length. The spores which due their huge dimensions belonging to the largest among fungi, develop in the asci in twos only.

Early morel grows from March to the beginning of May on humus-rich soil among fallen leaves in deciduous forests often it accompanies either aspens or rowans.

It is an edible mushroom with a very good flavour, which might perhaps be confused only with Morchella hybrida.

Discina Perlata

This is one of the characteristic species of Discomycetes whose fruit bodies appear soon after the snows melt and the forest soil has warmed up. Its apothecia are usually found on the cut surface or on the roots of spruce stumps overgrown with moss, often in glades and woodland clearings. They often grow gregariously and attract attention both by their patelliform shape and by their chestnut-brown disc with its wavy margin. A short, deeply furrowed white stipe grows on the underside of young fruit bodies, which makes them resemble the former genus *Acetabula* (today united with the genus *Helvella*). They differ from this by having completely different spores. The spores of *Discina perlata* are ornamented with a fine reticulum similar to that on the spores of *Gyromitra gigas*. Since the spores of both species are marked with small warts around each pole, they will be of no help in distinguishing these two fungi. *D. perlata* and *G. gigas* grow in similar habitats, in the same season of the year. Both species are edible and taste excellent.

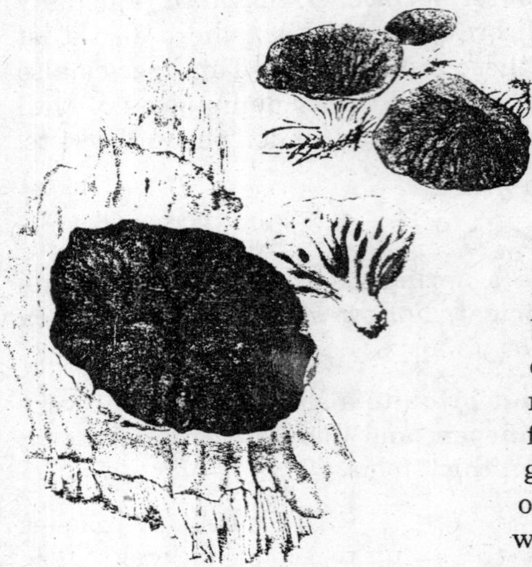

Rhizina undulata is very similar to *D. perlata* in shape and colour. It can easily be distinguished by its underside from which numerous perpendicular, tough, rootlike structures develop, attaching *R. undulata* to the substrate. It often appears in large quantities in summer and autumn on burned-over forest areas – especially on charred roots of pine-tree stumps, because its mycelium grows in pine wood. It is inedible

Summer Truffle, Common Truffle

True truffles are highly prized aromatic fungi, used since antiquity for culinary purposes. Since time immemorial they have belonged to the most widely known and most esteemed fungi.

Summer Truffle grows in some warm regions of central Europe, and also confined to oak forests.

Tuberous fruit bodies varying in size from 1 cm. up to 10 cm. Their development takes place out of sight, relatively deep undergound. Their gleba is whitish or yellowish at first, and interwoven with whitish veins, it is strongly aromatic at maturity one to six spores develop in each globular ascus; their surface is ornamented with a wide-meshed, angular, high reticulum. The fruit bodies grow from summer until winter.

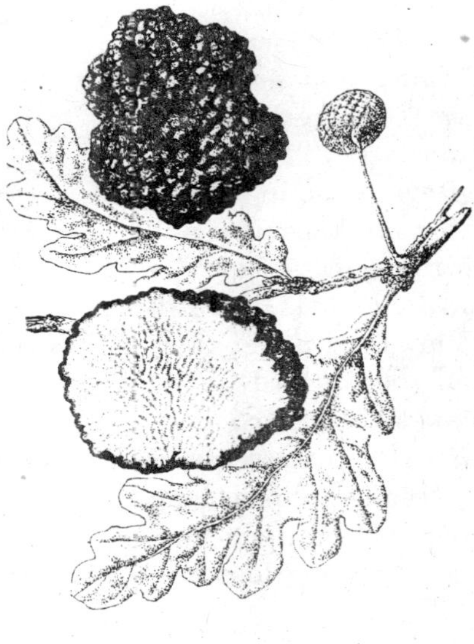

Although the Summer Truffle is not as pleasing to the palate as the Winter Truffle, or the Perigord Truffle it nevertheless belongs among high quality esculents. Of course the relatively coarse warts must be removed before their preparation for table use.

Otidea Onotica

Like the vast majority of other species belonging to this genus, *Otidea onotica* produces long fruit bodies that grow perpendicularly to the substrate. On one side they split lengthwise down to the base and have incurved margins, their attenuated basal part often tapers downwards into a short

stipe. Their colour is ochre, yellow or brown. The flesh is relatively thin and dry. They resemble hares' or rabbits' ears. The distinction between the individual species of the genus *Otidea,* several dozens of which are known, is far from easy because of their remarkable similarity.

Notwithstanding, *O. onotica* is one of the few easily distinguished species, differing from all the others in both size and colouring. On the outside it is cream-yellow to orange; the disc bears a pleasant pinkish hue. The fruit bodies are up to 13 cm. high and 1-3 cm. wide, which makes *O. onotica* one of the largest species in the family. It appears most frequently in small clusters but also solitarily in humus or in a deep layer of decaying foliage covering the floor of deciduous forests – usually under hazels, hornbeams and oaks, and less often in coniferous forests.

A characteristic of all the species of *Otidea* are paraphyses hooked at the apex and usually provided with a short excrescence, as well as broadly ellipsoid, smooth spores filled with two large droplets. The ascus membrane is nonamyloid.

Owing to their relatively rare occurrence and to the dry, thin flesh, the Otideaceae are of little significance for the practical mushroom gatherer.

Choiromyces Venosus

Many mushroom pickers confuse *Choiromyces venosus* with truffles. Its fruit bodies have an irregualr tuber-like shape; they are 5-15 cm. in size, very hard and relatively heavy. Their surface is off-white to greyish brown at first, gradually becoming yellow to brown. A cross-section reveals a tough inner matter (the 'gleba') shot through with a multitude of extremely thin, labyrinthine channels discernible as fine veins of unequal width and varying length. The gleba is white, with brown veins at the time when the spores are ripening. The asci arranged in the veins are clavate, bearing 2, 4 or 8 spores.

From the ecological point of view, *C. venosus* belongs to hypogeous fungi, as its fruit bodies start developing under the surface of the ground. It is only later that they partly jut out of it. *C. venosus* grows in summer and autumn in all types of woods, and in both sandy and heavier, clayey soils. It is widespread throughout the temperate zone of the northern hemisphere, even in regions where true truffles (genus *Tuber*) do not occur. It is an edible species but it is not advisable to keep ripe fruit bodies of *C. venosus* in a closed room owing to their penetrating smell which soon saturates the whole space and becomes unbearable. The fungus is used only in small doses as spice to season foods.

Krombholziella Aurantiaca

Boleti are very popular are those whose conspicuous red orange caps give colour to dark forest stands of aspen or birch.

Krombholziella Aurantiaca is associated with aspens, under which mushroom pickers, frequently search for it. The stipe is 6-15 cm. long and 1 to 3 cm. thick, covered in youth with whitish, in age with red brown to brown orange scales.

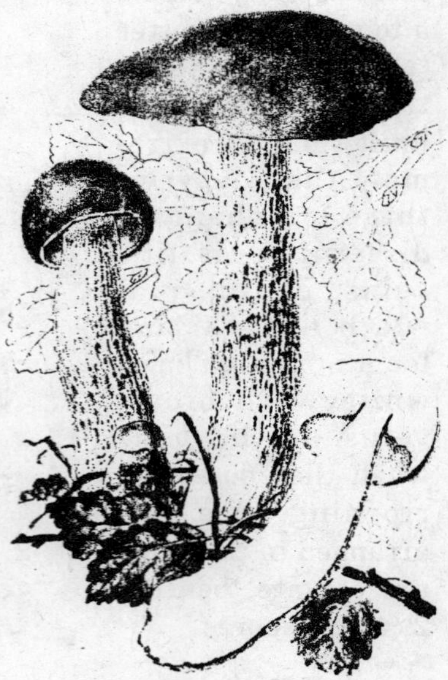

The cap is 4-15 cm. wide, dark brown-orange to red brown, pulvinate.

The tubes are whitish. The flesh is white, when cut it turns greyish pink at first then black. These are the essential features for distinguishing this species from the similar *K. rufescens*.

K. aurantiaca is an excellent and pleasant-looking edible mushroom; the fact that it is rarely infested by insect larvae is one of its advantages. It grows in summer and autumn from lowlands to the mountains in the whole north temperate zone.

Boletus Luridus

Boletus luridus is the second most widely known and also most commonly found blue bolete. The cap of *B. luridus* attains 5-20 cm. in diameter. Its cuticle is relatively pale in various light yellow or olive-yellow to brownish shades, often it is almost yellow or yellow-orange, but may also be pale carmine-red; it is finely felted, staining dark or blue when handled. The tubes are yellow, tinged olive-green or lemon-yellow when young, later yellow-green to dirty olive, with

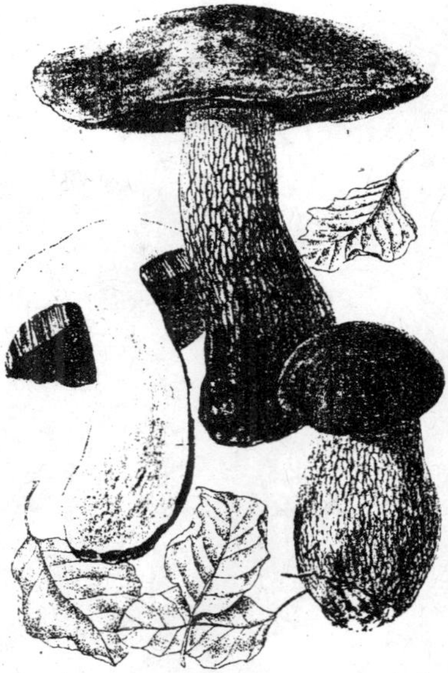

usually orange but also brick-red to blood-red pores. The oblong-clavate, 2-8 cm. thick stipe is yellow below the apex, in the middle and lower down the yellow colouring passes into red, near the base it becomes dark wine-red, even blakish brown in places. The stipe surface is covered all along its length by a coarse, red, long-meshed reticulum. The flesh is pale yellow, in the stipe usually wine-red; just like the tubes and the stipe surface, the cut flesh instantly turns ink-blue, while the red-coloured places stain violet. The flesh has a faint mushroom-like taste and smell.

This bolete, characterized by considerable colour variability, occurs predominantly in deciduous forests – mainly under oaks and beeches. It may also appear under conifers, mostly on a calcareous substrate and in warmer regions. It grows from May until the autumn. It is particularly the presence of reticulation on its stipe that constitutes the difference between the Red-stalked Boletus and the species discussed here. It is edible, yet the same applies to it as to the Red-stalked Boletus; the ingestion of raw or insufficiently cooked fruit bodies causes vomiting.

Boletus Aestivalis

Besides the Edible Boletus, *Boletus aestivalis* is the second best-known and most popular member of this genus. It is widespread in the whole of western and central Europe, and also occurs in North America and north Africa. Its favourite habitat are deciduous forests; it is far less common in mixed and coniferous forests. It can be found mainly under

oaks and beeches on whose roots it lives in mycorrhiza, but it is capable of establishing such a relationship with other deciduous trees as well. In Europe it can also form mycorrhizal associations – though rather exceptionally – with exotic trees kept in parks or planted in woods; thus, for instance, it was found under the North American oaks *Quercus rubra* and *Q. palustris*. However, finds have been reported even from established coniferous monocultures, from spruce, fir and pine forests where deciduous trees are entirely absent.

B. aestivalis grows from May until autumn. It differs from the related and very similar Edible Boletus by the structure and appearance of the cap cuticle, as well as by the surface of the stipe. The cuticle of *B. aestivalis* is matt, finely velvety to tomentose, lighter coloured, pale grey-brown, light, yellowish grey to leathery brown, often cracking in age. The stipe is uniformly pale brownish, marked with a coarse brownish reticulum. This reticulum either covers the whole stipe or reaches to more than half the length of the stipe.

Edible Boletus Cepe

This is the most popular mushroom growing wild, and is collected both for table use and for the purposes of large scale industrial production.

As a mycorrhizal fungus the Edible Boletus is associated primarily with spruces but also occurs in decidous forests though far more rarely.

Under specific conditions its mycelium is capable of forming mycorrhizal relationships also with oaks and beeches, eventually with other deciduous trees. This is infact no exceptional phenomenon or some species live in mycorrhiza with both deciduous and coniferous trees.

The range of the Edible Boletus is extremely variable and there are different views about the classification of individual variations.

In its typical form the Edible Boletus, has a deep brown to black brown, smooth and glossy cap cuticle, slightly felted only in youth.

It is always somewhat darker than in the related Boletus aestivalis. At maturity the stipe is white only near the apex, while in its lower parts it is brownish.

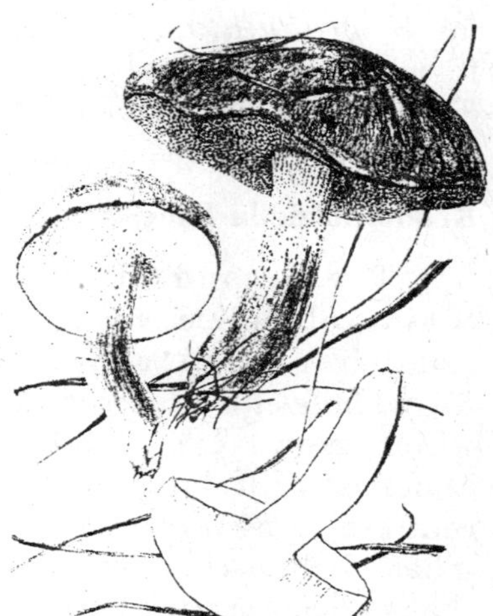

Suillus Granulatus

Suillus granulatus has no veil, consequently no ring is present on its stipe. This species strongly resembles *S. placidus* from which it may be distinguished by the absence of the ring as well as by a lighter-coloured cap with always some shade of yellow-red, drying to ochre-yellow. The relatively short tubes are often decurrent, pale yellow or dirty yellow at first, ochre yellow to brownish yellow in age. A conspicuous feature of fresh fruit bodies are plentiful droplets of a milk-coloured liquid oozing from the tube pores. They adhere to the surface of the stipe immediately below the tubes, drying brown and changing into brown or reddish granules. The stipe is yellowish, later brownish, with a white mycelium at the base. The whitish to yellowish context rapidly becomes soft and watery and has a slightly sour, fruity taste and smell.

S. granulatus has developed a mycorrhizal relationship with pines having their needles arranged in pairs – primarily with the Scots Pine, Mountain Pines, and the European Black Pine. It occurs under these trees from lowlands up to the mountains; for example in the Alps it is capable of growing at an altitude of 2,250 m. It is often gregarious in grass along woodland paths or in the outskirts of pine woods and shows a preference four lime-rich soils. It is an excellent ediable mushroom.

Sporadically we may also come across ringless forms of *S. luteus* which could easily be mistaken for *S. granulatus*. These forms, however, always differ in the colour of the cap, pores and stipe.

Krombholziella Rufescens

The second of the couple of boleti with a reddish orange cap is *Krombholziella rufescens,* which grows exclusively under or near birches. In contrast to *K. aurantiaca,* it has a lighter-coloured cap that is deep orange-brown to yellow-brown. The cap is 5-15 cm. wide, pulvinate, usually with a membranous border. The stipe is 6-15 cm. long and 1-3 cm. thick, the minute scales covering it are black or brown-black. The tubes are tinged grey already when young. When injured or cut, the originally whitish flesh slowly turns grey-violet with a pink flush, and greenish blue in the basal portion of the stipe. This colouring, however, is not an exclusive characteristic, since it also sometimes appears in *K. aurantiaca.*

Like *K. aurantiaca,* this species is also edible, very tasty, and suitable for all culinary purposes. It grows in summer and autumn in most countries of the north temperate zone where it shares the habitat of the birch.

Some closely related species grow under other trees, e.g. *Leccinum vulpinum* under pines and *L. piceinum* under spruces. Characteristic of both these species is a rusty brownish orange cap whose colouring resembles that of *Lactarius rufus.* The cap of *Leccinum quercinum,* growing under oaks, has a more pronounced whitish shade.

Boletus Aereus

This beautiful bolete is an explicitly thermophilous species. It accompanies old oak and other deciduous forests situated without exception in protected warm localities with thermophilous flora. It is more common in southern Europe.

It is characterized by a dark chocolate to black-brown, 5-20 cm. wide cap bearing a dry finely velvety cuticle becoming later glabrous and sometimes breaking up into irregular patches.

In sharp contrast with the dark colour of the cap are the white tube mouths, turning green-yellow with an olive tinge upon maturity.

The stipe is substantially lighter-coloured than the cap covered with a fine reticulum on the surface, the flesh is white only under the cap cuticle, is light flesh-brownish sometimes staining slightly pink when cut. It is relatively firm and softens only with age. It has a mild taste and almost no odour when fresh, but when dried it emits a pleasant mushroom-like smell.

It appears as early as May in pine forests. These species are among the best edible mushrooms.

Suillus Viscidus

In older mycological literature *Suillus viscidus* is usually described under the name *Boletus aeruginascens*. Its cap is 4-13 cm. in diameter, with a slimy and glossy surface, greyish white, yellow-grey, olive-grey or grey with a reddish flush, sometimes with darker striae, fibrillose-scaly when dry. In young fruit bodies the cap is connected with the stipe by a whitish or yellowish membranous veil whose remnants may be found at the cap margin of mature specimens. The tubes are whitish, then dark grey to brown-gey, with relatively large, angular pores passing over to the stipe apex in a retiform arrangement. The 6-8 cm. long and 1.5-2 cm. thick stipe is rather thin, yellow-grey, with a transitory ring. The stipe surface under the ring is covered with irregular pits. The flesh is white, later greyish, yellowish or brownish in the stipe, readily turning blue when cut above the tubes. It has an inconspicuous smell and a pleasantly sour, fruity taste.

This species grows in a mycorrhizal relationships with various larch species; in Europe it is the European Larch (*Larix decidua*) which *S. viscidus* accompanies high up into the mountains, to an altitude of 2,300 m. It occurs from May to October, ranging throughout the north temperate zone where it occurs both in established larch stands and under planted and cultivated larches. Although it is edible and relatively savoury, mushroom pickers usually prefer to bypass it due to its rather unattractive colouring.

In the Alps and in the High Tatras, the variety *bresadolae* has been found. It has a darker, red-brown cap and stipe, lilac rosy pores and a lemon-yellow flesh.

Chalciporus Piperatus

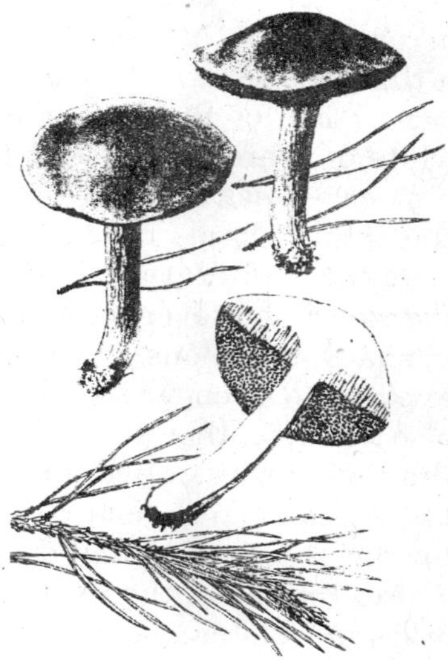

The genus chalciporus combines the qualities of the genera Xerocomus and suillus.

Its fruit bodies are small, the cap cuticle is dry or slightly slimy, glossy when dry, the stipe is thin the flesh and the mycelium are yellow.

It forms a mycorrhizal association with both coniferous, and deciduous trees.

It has the tough stipe, which is 3 to 6 cm. long, and 0.5 to 1 cm. thick often curved and narrowing towards the base. It is the same colour as the cap bright yellow at the base.

The flesh is whitish-yellowish in the cap, and yellow in the stipe.

It has a hot peppery taste and is almost odourless.

C. Piperatus grows from July to November, being most common in pine woods. In deciduous forests it particularly favours, beech and hornbeam stands and is also found under birches.

These mushrooms are edible only when mixed with other mushrooms.

Xerocomus Porosporus

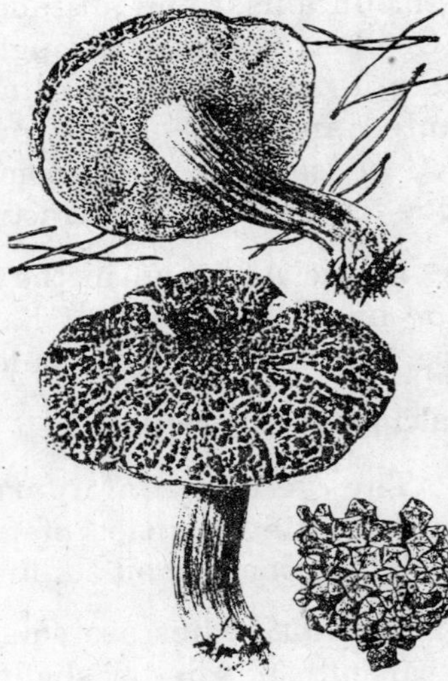

Species related to the Red-cracked Boletus also include *X. porosporus*. European mycologists treated it for a long time as *X. truncatus,* irrespective of the fact that *X. truncatus* is a different. North American species. The only macroscopic difference between *X. porosporus* and *X. chrysenteron* is that the chinks in the cracked cap cuticle of the former are white or ochre-yellow, and not pink or reddish as is the case in the typical forms of the Red-cracked Boletus. This difference, however, primarily relates to fruit bodies growing in shady conditions. In fruit bodies of the Red-cracked Boletus exposed to stronger sunshine the cracks in the cap cuticle are also white, for they penetrate deep into the cortex; nevertheless, the rosy flush is discernible on at least parts of the cap. The two species can be reliably distinguished under the microscope by the shape of spores. In *X. porosporus* the major part of spores (not all of them) are truncate, while spores of the Red-cracked Boletus are always rounded at the top. There is no difference between the mentioned species in any other characteristic. As *X. porosporus* remained unidentified for a long time, the data about its range of occurrence are still incomplete. It certainly grows in several European countries and in North America, mostly in mixed forests under oaks, hornbeams, aspens, spruces and pines. It is edible.

Xerocomus Rubellus

In 1836 the Prague physician and mycologist V.J. Krombholz described a bolete found near Prague greatly resembling the fairly common *X. chrysenteron* but distinguished from it by its conspicuously red-coloured cap. In the course of time this mushroom was repeatedly described under various names and in various countries of Europe and North America.

With a view to the date of description however, Krombholz's designation is given priority.

X. rebellus grows predominantly under deciduous trees, mostly, oaks, limes, hornbeams, and beeches, especially at lower elevations, in sunny warm places, frequently accompanying tree rows along roadsides; its occurrence is typically solitary and rather rare, otherwise everything applying to *X. chrysenteron* also holds good about this species.

Bay Boletus

The Bay Boletus has a chestnut- or chocolate-brown, smooth, 5-12 cm. wide cap, slightly sticky in wet weather. The tubes are pallid, later yellowish to yellow-green, turning green after handling, and the whitish flesh turns slightly blue when cut. The cylindrical stipe, frequently hooked and narrowed towards the base, is pale brownish, 3-8 cm. long, 1.5-3 cm. thick. The flesh has a pleasant mushroom-like smell and taste.

It grows profusely in coniferous forests, mainly in submontane regions, from summer to late autumn. Sometimes it grows directly from the foot of trees or on rotten stumps, not uncommonly also out of decayed spruce cones into which its mycelium has forced its way from the ground. It often appears in woodland margins where its fruit bodies are frequently hidden in low grass. There are regions where it proliferates in spruce and pine forests.

It is gathered as a tasty edible mushroom. The fact that it grows until late autumn and that it is rarely attacked by insect larvae is an advantage. However, the mushroom is a favourite food of rodents, thus it often happens that its fruit bodies bear traces of their teeth. It cannot be confused with any other similar species, perhaps only with the Bitter Boletus which is also abundant in coniferous forests. This, however, has a much lighter-coloured cap and a markedly reticulate stipe, pinkish tubes, and its flesh, not staining blue when cut, is repulsively bitter.

Phylloporus Rhodoxanthus

Phylloporus rhodoxanthus is the only European representative of its genus. Originally it was described as a gill fungus under the name Agaricus pelletieri, later it was considered to belong to the genera clitocybe or paxillus.

Its actual place however is among the boleti. The erroneous assumption that this is a gill fungus was substantiated by the peculiar structure of the hymenophore which infact assumes the shape of thick gills inter-connected by numerous transverse ridges of various height forming variously sized retiform cavities in the basal part of the gills.

The cap of the fungus has, beautiful golden yellow hymenophore. The cap is 3 to 7 cm. wide, olive to purplish-brown. The cylindrical stipe is 3 to 5 cm. long and 1 to 1.5 cm. thick narrowing downwards, often ecentric reddish or yellow-brown. The spores show its close relation to the boleti.

This interesting mushroom grows solitarily in no particular abundance, in deciduous mixed and coniferous forests, most frequently along forest paths, in clearings among grass, woodland ditches etc.

Xerocomus Parasiticus

Xerocomus parasiticus represents one of the sporadic instances of parasitism among boleti. Furthermore, it is a noteworthy species from the point of view of its ecological and phytogeographical distribution. It parasitizes and partly destroys the fruit bodies of *Scleroderma citrinum* and *S. polyrhizon*. As a rule, the fruit bodies of *X. parasiticus* grow in small clumps out of the basal part of the tuber of *Scleroderma* fruit bodies. Although the host is very common and plentiful, its parasite is rather rare. It has been reported from most European countries and from north Africa. It resembles a small *X. subtomentosus* in size, appearance and colouring. Its cap is 2.5-7 cm. wide, subglobular in youth, expanded in age, leathery yellow or yellow-brown, changing to olivaceous, finely felted, slightly cracking round the centre when the weather is dry. The tubes are pale yellow, later lemon-yellow to olive-brown, and adnate. The stipe is cylindrical, rather thin, tapering downwards and hooked, pale yellow, brownish to olive-brown and finely dotted. The light yellow flesh is more deeply coloured in the stipe and above the tubes. Its taste is mild and its smell inconspicuous.

Due to the extreme rarity of its occurrence and the small size of fruit bodies, it is of no significance for practical mushroom pickers – on the contrary, its protection is most important.

Suillus Placidus

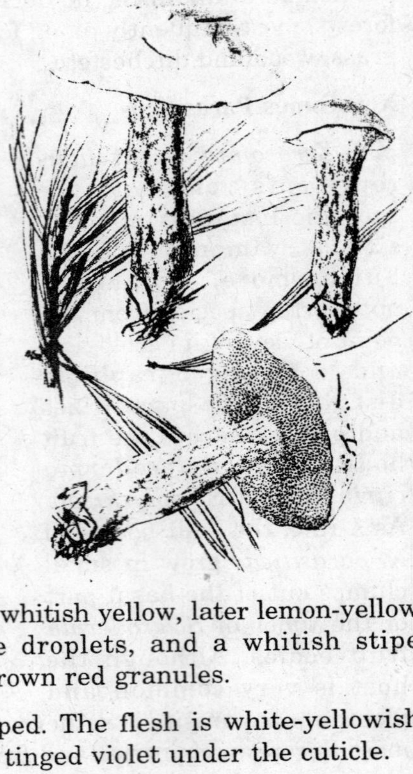

All species of Suillus are mycorrhizally dependent on conifers, some of them specializing in particular genera or even in particular species.

The fruit bodies of *S. Placidus* appear most often gregariously from June to October. They differ from most of the other species of suillus in their whitish or yellowish white cap, often with a lemon-yellow tinge, which in older fruit bodies sometimes passes over to a shade of violet. They have whitish yellow, later lemon-yellow tubes often exuding white droplets, and a whitish stipe covered with red-violet or brown red granules.

The veil is not developed. The flesh is white-yellowish above the tubes, sometimes tinged violet under the cuticle.

Suillus Placidus is an edible mushroom.

Gyroporus Castaneus

This neat bolete is by no means a common species. The 3-8 cm. wide cap is slightly convex, soon flatly expanded, red-brown to chestnut-brown, with an almost velvety-tomentose surface, dry. The stipe is 8 cm. long and 3 cm. thick, the same colour as the cap; it is also finely tomentosevelvety, frequently striated and wrinkled, with a surface that often appears crumpled. Its context, white in youth, is soon partitioned into

chambers or it becomes hollow altogether. The veil is absent. The tubes are white at first, then cream yellow, almost light lemon-yellow at maturity, often staining brown when handled. The tube pores are relatively minute and the same colour as the tubes. The mildly tasting flesh is almost without odour. The spore print is pale yellowish.

The favourite habitat of *G. castaneus* are dry, warm and sparse woods on acid soils where it gives preference to open places exposed to sunlight. Sporadically it occurs also outside the forest in ditches, lanes, field margins and pastures. It grows from July to November and is widespread in the warmer regions of the entire north temperate zone. Though edible, its taste is not particularly appetizing.

G. cyanescens, the second European representative of this genus, is characterized by its flesh and tubes staining deep blue when cut or handled, and by a straw-yellow or yellow-brownish cap. It occurs in sandy pine forests or in mountain beech forests. It is also edible.

Strobilomyces Floccopus

This peculiar bolete evokes a somewhat gloomy impression. Characteristic felt-like scales cover the whole surface of its 5-12 cm. wide, umber-or sepia-brown cap. The large recurved scales are blackish at the top and almost erect near the cap margin. The woolly veil covering the tubes at youth leaves appendiculate fragments along the cap margin

of fully developed fruit bodies. The whitish tubes as well as their relatively large angular pores gradually turn grey to brown, staining red where bruised. A fluffy annulus is clearly discernible on the cylindrical stipe. The flesh is whitish when young; in fresh fruit bodies it turns pale orange when cut, then dirty violet. Old fruit bodies have a markedly dry, cottony consistency, a mild taste, and an indefinable smell. The spore print is almost black.

Strobilomyces floccopus occurs in coniferous and mixed forests, and only rarely in pure deciduous woods, being most abundant under conifers on noncalcareous soils. As a general rule it grows solitarily, usually in summer and in autumn. In Europe it is the only representative of its genus; it is distributed throughout the temperate zone of the northern hemisphere. *S. floccopus* is edible; however, its general appearance and the excessively dry flesh of older fruit bodies do not contribute to its attractiveness as an edible mushroom.

Lactarius Volemus

The attractive fungus well-known to practical mushroom pickers grows in dry summer weather when other edible mushrooms are rather scarce. Recently it has been disappearing from European forests – just like a number of other fungi – and there are regions where it has become rarity, although it used to be quite common there only 30 years ago.

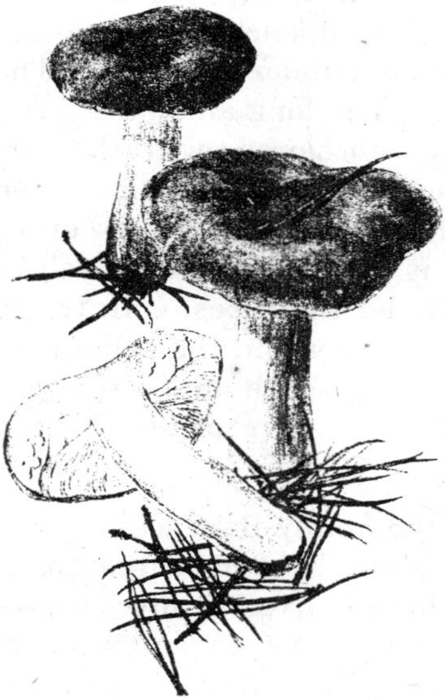

One of the reasons of this phenomenon is an ever increasing exploitation of forests accompanied by intensive wood production systematically decimating old forest stands, including pine woods.

This fungus has, brownish orange cap, about 5 cm. to 15 cm. wide.

It has glabrous, pruinose cuticle, a large amount of pure white, sweet latex oozing from the injured flesh of fresh fruit bodies.

Sometimes Lactarius volemus is mistaken for *L. mitissimus,* a small reddish orange species which among other things, is distinguished by its acridly tasting latex its flesh does not turn brown and there is no smell of trimethylamine. This mushroom is edible.

Gyrodon Lividus

The genus *Gyrodon,* monotypical (i.e., including one species only) in Europe, differs from other boleti by having extraordi-narily short tubes. They are only 2-5 mm. deep and not easily detachable from the cap context. Their relatively large pores are irregularly sinuous, radially arranged, and clearly decurrent.

Gyrodon lividus has a 4-10 cm. wide cap, whitish with an ochre shade, then rusty brown or olive-chestnut, clearly felted, sticky when moist, turning blue under the cuticle. The stipe is usually thin and short (3-12 cm. long and 0.5-2.5 cm. thick), often curved and eccentric, concolorous with the cap. The tubes and pores are at first lemon- to golden-yellow, then green-yellow and olive-brown, discolouring bluish green or brown when handled. The flesh is pale yellow, brownish at the base of the stipe, deep blue under the tubes, with traces of greyish blue and then of reddish brown in the remaining parts of the fruit body. It is softly spongy, with an indistinctive smell and a somewhat acid taste. The spore print is ochre-brownish.

This interesting member of the family Boletaceae grows on wet, often grassy, ground abounding in sedge tufts, exclusively under alders with which it lives in mycorrhizal association. It prefers calcareous substrates and grows from July to October. Its range covers the whole of Europe, extending eastwards beyond the Urals. It is a rather rare species which, though edible, is of little significance for practical mushroom pickers due to its sporadic occurrence and the excessively soft consistency of the fruit bodies.

Porphyrellus Pseudoscaber

This is, European representative of the genus Porphyrellus greatly resembles strobilomyces floccopus in colour, but possesses neither scales nor a veil. The spores are also different.

The whole mushroom has a sooty appearance. The 5-16 cm. wide cap is pale grey at first, gradually darkening into various shades of olive to umber brown, staining almost blackish when handled.

Its cuticle is dry almost felted to velvety, and is not capable of being peled off, also the tubes grey in youth turn darker on handling, but sometimes they stain slightly bluish or even reddish. Their pores are angular with a denticulate mouth.

The surface of the cylindrical or claviform stipe, frequently hooked at the base and of the same colour as the cap surface is finely felted-punctuate and longitudinally striate. When cut the pallid flesh stains bluish green, or greyish red at first, then brown to black.

P. Pseudoscaber occurs most abundantly in coniferous and mixed submontane forests, being less common in deciduous forests.

Though edible its taste is less appetizing than that of other boletes.

Slippery Jack

This species grows exclusively under pine species whose needles are arranged in pairs. In Europe this is primarily the Scots Pine and the Mountain Pine (*Pinus mugo mughus* and *P. mugo rotundata*). It is only very rarely that the Slippery Jack appears under the European Black Pine and several alien species kept in parks and gardens. It occurs abundantly within the range of the mentioned pines which it accompanies up to the northernmost limit of their distribution, e.g. to northern Siberia and the Kola Peninsula.

The 4-12 cm. wide cap has a chocolate- or yellow-brown, readily peeling cuticle, slimy and glossy when moist, radially fibrillose when dry. The tubes are pale yellow, then lemon- to golden-yellow, with relatively small pores. The stipe is cylindrical, lemon-yellow, strewn with yellow or brown granules above, brownish below, sheathed with a conspicuous tomentose-leathery veil. The flesh is whitish, then lemon-yellowish, with a pleasant fruity taste and smell.

It is the veil, which leaves an annulus on the stipe and sometimes persists at the cap margin in the form of scattered patches, that differentiates the Slippery Jack from the otherwise very similar *S. granulatus*. The latter is an excellent tasting edible mushroom.

S. bovinus has no veil either; it has a light ochre-brown cap with only a partly separable cuticle, the stipe is thin and finely fibrillose. It grows under pines on sandy soils. It is edible but of an inferior quality.

Brown Birch Boletus

In forests and their outskirts, it is birches that often hold the attention of mushroom pickers. Underneath, they most frequently seek members of the genera *Krombholziella* and *Leccinum*. The Brown Birch Boletus is in fact the best-known representative of Boletaceae whose mycelium lives in a symbiotic relationship with birch roots. Fruit bodies of the Brown Birch Boletus appear in summer and autumn both under old birches and under young trees in clearings and copses all over the north temperate zone. Its grey to brownish, 5-15 cm. wide cap is seated on a relatively thin and long (6-15 × 1-1.5 cm.), white stipe covered with almost black scales. The flesh is white or whitish, almost unchanging when cut, drying black. It has a pleasant mushroom-like taste and smell.

The Brown Birch Boletus is a tasty edible mushroom, but has one disadvantage; its cap soon becomes soft. On the other hand, its stipe is excessively tough to fibrous-woody. This is why young mushrooms are most suitable for collection. Even though the flesh turns black in the process of cooking, this does not affect the taste.

Closely related to the Brown Birch Boletus are several similar species. These differ in the species of woody plant they depend on, as well as in the colour reaction of the bruised context. *K. holopoda* is found – though rather rarely – under birches on peaty soils. It used to be considered an albino form of the Brown Birch Boletus. Most similar to the latter in appearance is *Leccinum carpini* with its dark brown to nearly black, often pitted cap; the whole fruit body is rather tough, the flesh turns grey-black on injury. It grows sporadically in hornbeam woods.

Xerocomus Subtomentosus

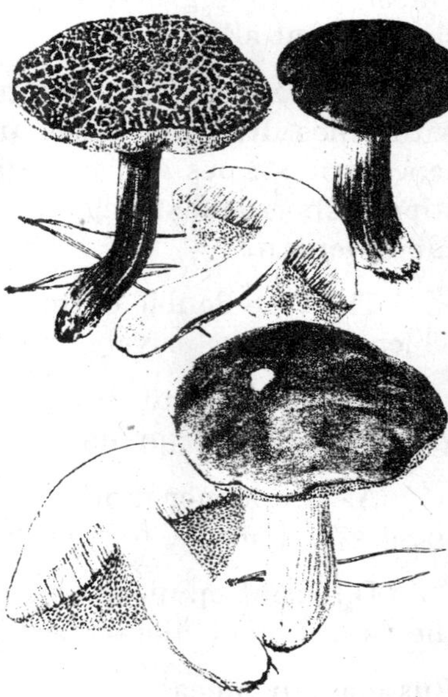

X. subtomentogus is more highly appreciated by mushroom pickers. These species have wide cap without, any cracks; the cortex layer underneath the cuticle does not contain any purplish red pigment, the tubes are of a nicely rich chromium yellow, the stipe is usually ribbed near the apex and narrowing downwords.

The red cracked Boletus closely resembles, the X. subtomentosus.

The tubes of X. subtomentosus are pale yellow, sometimes green yellow, the stipe is cylindrical, mostly not ribbed, red tinged in places, ordinarily not attenuated at the base. It is an edible mushroom.

Saffron Milk Cap

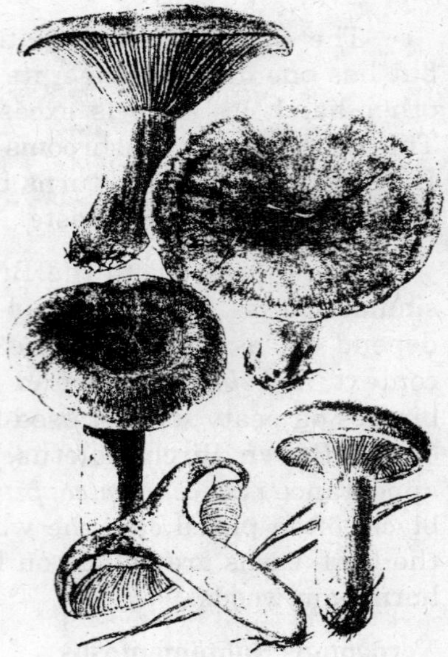

This fungus has the fruit bodies of shed bright orange or wine-red, latex where injured.

The best example of this Saffron Milk Cap commonly gathered as a tasty edible mushroom. Its latex is carrot red when fresh, it does not change colour later at most it dries to a paler shade.

Its light orange-ochre cap turns almost ochre in places turning green only slightly if at all.

The Saffron Milk Cap grows in mycorrhizal association with pines. Its fruit bodies are, sturdy its cap bears distinct concentric stripes and may grow to 20 cm. in diameter. The stipe is short and strong, never exceeding the diameter of the cap in length.

The true double of the saffron milk cap is *L. deterrimus* which grows exclusively under spruces.

Its slender fruit bodies usually have a long stipe and a green banded cap, green spots develop also on the gills.

After a longer time interval its milk changes colour from the original orange to wine-red.

Different opinions also prevail among mycologists as to the definition of characteristic features of *L. deliciosus*.

Russula Sanguinea

When strolling through pine woods in late summer or autumn, we may find the beautifully coloured and shapely *Russula sanguinea* growing by moist, grassy forest paths. Its

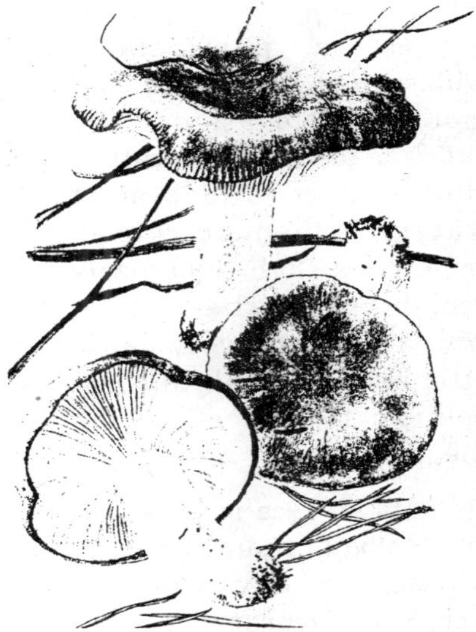

cap grows up to 10 cm. in diameter; it is moderately fleshy, soon flat, broadly depressed, narrowing towards the margin, incurved, and often wavy. Its cuticle is blood-red or carmine, eventually rosy, discolouring in places to pale cream or even whitish. In rainy weather the cuticle is somewhat sticky, otherwise it is dry and matt, undetachable. The stipe is relatively short, tapering towards the base, red or pick for the most part, but often marked with lemon-yellow spots; the flesh also shows a tendency to turn yellow. The obviously decurrent gills are whitish when young, later they become dark cream-coloured, sometimes with reddish edges. The flesh is white, pink under the cap cuticle near the margin, with a slightly acrid fruity smell. The taste is usually rather pungent, sometimes bitter, exceptionally almost mild.

The taste of the flesh in many red-coloured russulas is acrid, though to a varying degree. Probably the most acridly tasting russula is *R. badia* whose flesh smells strongly of cedar wood. After tasting it a burning feeling remains on the tongue for a long time, sometimes resulting in a loss of sensitivity. The chemical composition of resinaceous substances which are assumed to cause vomiting and diarrhoea after the ingestion of acrid russulas is still unknown.

Megacollybia Platyphylla

This robust fleshy mushroom grows solitarily in summer and in autumn in deep, damp especially coniferous forests on stumps or in the humus arisen from decomposing branches and roots.

It is particularly widespread in foothill and mountain elevations. The cap is 5-15 cm. wide, rather thin fleshed and almost watery, expanded-convex sometimes slightly obtusely umbonate in the centre, with a long incurved, sharp, attenuated, membranously pellucid margin exceeding the gills.

Its surface is dark ash grey smooth, radially darker fibrous in dry weather fibrously lacerated over a white ground colour. The stipe is often shorter than the diameter of the cap, which is tough elastic, coarsely fibrose, grooved cylindrical, light grey to whitish, its base is overgrown with numerous tough, white mycelial threads penetrating through rotten wood.

The gills are very distant, and broad, whitish with often uneven, brown-rimmed edges. The spore print is pure white. The flesh has neither smell nor taste. But it is edible.

Lepista Nebularis

Lepista nebularis is one of the most easily identifiable of gills fungi. Its robust stature, appearance, colouring and specific odour are properties characterizing it so strongly that it cannot be mistaken for any other species. Also its relatively late occurrence in autumn is typical. It appear—in clumps as a rule—from September until the first frosts, usually in rather deep forest humus among fallen leaves and needles, but sometimes also in parks and gardens. The fruit bodies last for a long time in their habitat. The cap sometimes exceeds 20 cm. in width (usually measuring 8-12 cm.); it is ash-grey or greyish brown, whitish pruinose when young, dry, fleshy, and convex. The stipe is markedly clavately expanded towards the base, fibrous, greyish, and firm-fleshed. The originally white gills gradually shade to yellow: this is called forth by the colour of the spore mass. The flesh is white, unchanging, sweetish-acid, with a quite specific aroma—according to some authors it has a farinaceous smell, others compare it to that of orange blossoms, but what is involved here is most probably a mixture of both these scents with a mouldy tinge.

Although *L. nebularis* is edible and popular with a number of mushroom pickers, it is much too aromatic and, in some cases, causes vomiting or diarrhoea. Recently it has been found to contain organic substances of the antibiotic type. Until lately it was currently classed into the genus *Clitocybe* from which it was excluded only a short time ago.

Stropharia Rugosoannulata

For the time being this is the only mushroom that can successfully be grown in one's own garden. Its cultivation is much simpler than that of field agarics.

Straw of various cereals or flax stems are used as a living substrate. This corresponds to its natural substrate, since in the wild it occurs on rotten straw. It grows from June to November.

Originally it was described from North America and later from Italy, but the idea that both might be just two geographical races of the same species cannot be ruled out. In the course of cultivation strains with diversely coloured caps were obtained; of these the highly productive cultivar 'vinnetou' with a pale red-brown cap grey-blue gills and yellowish white stipe is raised most commonly.

The typical form occurring in the wild has a cap 8-15 cm. in width, at first grey-brown, dingy yellow brown, growing paler later on and very sticky. The stipe bears a double ring, white and grooved above yellowish white below.

The flesh has a faint raddish-like smell and a mild taste, and is relatively tough. The spore print is purple brown to violet black-brown.

Common Funnel Cap

In a number of manuals this *Clitocybe* is introduced and described under the name *C. infundibuliformis*. It has a 3-8 cm. wide, funnel-shaped cap, bluntly subumbonate in the centre, relatively very thin-fleshed and rather elastic, with a quite typically scalloped-striate margin. The cap cuticle is fibrous, finely felted, almost unicoloured – pale ochre-yellow, light buff or creamy. The stipe is slender, clavate, approximately of the same length as the diameter of the cap, 0.5-1 cm. thick, whitish, enveloped at the base in a felted coating of white mycelium. The gills are crowded, deeply decurrent, and white. The context is creamy,

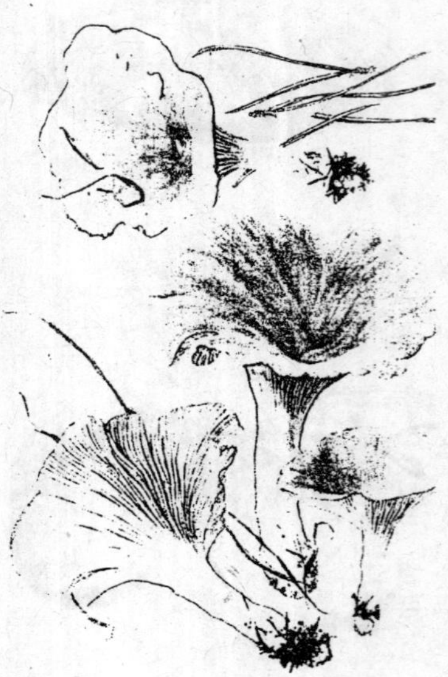

with a pleasant but faint smell of bitter almonds and a mild taste. The spore print a white.

The Common Funnel Cap grows profusely in coniferous forests, most often in moss and grass, but also among shed needles, usually in small groups. It is very similar to and is still confused with *C. incilis,* distinguished by a non-umbonate cap usually with a wavy and smooth, non-striate margin, greyish cinnamon gills, a slightly urinous smell, and also by somewhat larger spores. The question remains open whether these two species are actually independent: whether in fact they do not constitute a single species. These *Clitocybe* species are edible and collected in some countries as 'forest Fairy-ring Champignons'.

Clitocybe Odora

The colours prevailing in fungi of the genus *Clitocybe* are white, whitish grey, greyish brown and brown. Only a very few species are coloured differently, and *Clitocybe odora* is one of them. Not only the verdigris blue-green to green-grey colouring of the cap, fading to off-white, and the green-tinged gills and stipe, but also the pleasant sweetish aniseed scent of the whole fruit body make the confusion of this fungus with another species practically impossible. Its smell can also be compared to fennel. There are two other *Clitocybe* species emitting a similar scent – *C. suaveolens* and *C. fragrans.* Both of them, however, are quite differently coloured: their cap is pale greyish to ochre-brownish when moist, white when dry (i.e. the cap is typically hygrophanous).

C. odora has a 3-10 cm. wide cap, umbonate at first,

then broadly funnel-shaped, smooth and silky. The stipe is 3-8 cm. high and 0.6-1.2 cm. thick. In calm weather the smell of *C. odora* spreads to a distance of several metres from the fruit bodies, particularly if more of them grow close together. It is found in autumn, mostly in spruce woods among decaying needles. It is edible, and is best suited for flavouring less aromatic mushroom mixtures. Drying makes it lose its typical odour.

Red-stalked Boletus

The Red-stalked Boletus is a typical representative of Boleti whose flesh instantly turns dark blue on exposure to air. The bluing of the flesh is best seen in fresh fruit bodies and represents a characteristic common to several species differing in the colour of the cap and stipe surface.

The cap of the Red-stalked Boletus can be 20 cm. wide, however it is smaller, most often dark-brown with a felted to velvety and dry cuticle turning almost black in bruised places.

The tubes are 2-3 cm. deep, yellow-green with relatively small pores which are yellow only when young.

Soon they acquire a vermillion to dark blood-red colour weathering to orange in age. The stipe is clavate cylindrical, 2-4 cm. thick more intensely yellow above, red tomentose-punctate to scaly on a yellowish background in the middle, tinged olive-green at the base. The flesh is deep yellow, quickly changing to deep blue when handled or cut. The same applies to tube pores and to the stipe surface. It has a faint mushroom-like smell and a mild taste. The Red-stalked Boletus grows mainly in spruce forests. It is edible.

Hygrophorus Marzuolus

Hygrophorus marzuolus is the opposite of the Pinewood Wax Cap. Its robust fruit bodies start pushing themselves through the forest humus in early spring, as soon as the first rays of sunlight begin to warm up the earth. Sometimes they make their appearance at a time when there are still patches of snow. The first fruit bodies grow as early as February, yet the main period of growth sets is in April, exceptionally – after long-lasting and severe winters – in May and June. The fruit bodies may reach the size of 12 cm. in diameter and have a thick layer of firm flesh; at first they are white all over but soon the cap surface turns pearl-grey or black-grey, the cuticle is matt, very finely radially fibrous. The fruit bodies are often rather deformed and contaminated by soil, since they grow from a

greater depth. Groups of fruit bodies develop under the surface of the earth which they heave up like mole hills, thus betraying the place of their growth. The short and thick stipe tapers downwards, white with a greyish tinge, fibrous, the gills are thick, distant, shortly decurrent, white, later slightly bluish grey to pale grey. The firm white flesh has a sweetish smell and a mild taste. The whole surface of the fruit body is dry. The spore print is white.

Particularly interesting is the geographical distribution of *H. marzuolus:* it has been reported from Europe only, being most abundant in the Alpine region, the northernmost limit of its range is probably in Bohemia. It is most profuse in established coniferous or mixed stands, being rather scarce in deciduous forests (here it particularly favours oak stands). It is an edible, nutritious mushroom.

Chroogomphus Rutilus

Chroogomphus rutilus is another commonly occurring member of this group. It grows in summer and autumn in coniferous forests, particularly in pine woods. Unlike *Gomphidius glutinosus,* it is only slightly viscid or only sticky when moist. It has a 5-13 cm. wide, yellow-brown to orange-yellow umbonate cap and deeply decurrent yellowish to yellow-orange gills. When mature, the gills blacken by the spore mass. The stipe is 5-10 cm. high and 0.5-1 cm. thick, the same colour as the cap, with a fugacious superior ring. The flesh is soft, orange-yellow. An interesting microchemical reaction can be observed on the context of fresh fruit bodies: a drop of soda solution instantly changes the colour of the context to violet, whereas when touched with green-vitriol solution it stains vivid green, and in an aqueous ammonia solution it stains carmine. The cuticle of both cap

and stipe change to pink, together with the flesh, when the fruit body is drying up.

Like *Gomphidius glutinosus, C. rutilus* is also an excellent edible mushroom suitable for many culinary purposes. It turns violet when cooked and almost black when pickled in vinegar – but this does not lower its quality.

Common Ink Cap

Fruit bodies of the Common Ink Cap usually grow in clumps from rotten wood, concealed in the ground, and thus seem to be growing, directly from the ground.

They mostly occur in fields, on rubbish heaps in gardens, along roads, sometimes on compost heaps, and may be found from spring until autumn, especially after heavy rains.

The cap is egg-shaped, at first 3-7 cm. high later bell shaped or conical, 4-10 cm. wide with a wavy-lobed margin upturned when old.

The original ash-grey or greyish brown colouring of the cap turns black as the spores gradually mature.

The colour change proceeds from the margin to the cap centre. Ripe fruit bodies dissolve into a black pulp.

The cap cuticle is brownish and scaly, later glabrous. The gills are crowded, free, whitish when young. The stipe is 7-18 cm. long and 1 to 1.5 cm. thick, cylindrical white, finely fibrous, with silky sheen, at first full then hollow, ringless.

The flesh is odourless, with a mild taste. The spore print is dark brown. Young fruit bodies are edible and tasty.

Gomphidius Roseus

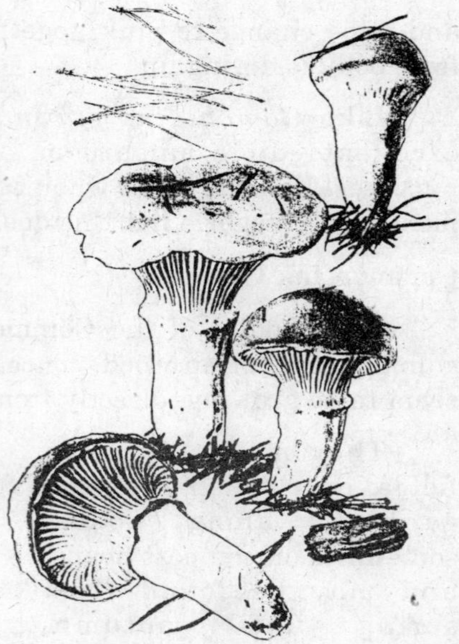

This is the smallest member of the genus *Gomphidius* which can hardly be confused with any other species. Its cap is 3-6 cm. wide, and of a beautifully light pink or greyish pink; more rarely it is brick-reddish in colour. This colouring weathers in age to some shade of brown, and black stains appear on the cap surface. When moist, the cap cuticle is slightly slimy, in young specimens it is attached to the stipe by a white veil. Young gills are white, then grey-brownish to almost black, distant, thick, and decurrent. The cylindrical white stipe, narrowing downwards, is tinged with brown or black at the base. The slimy ring soon disappears from its surface. The context is white, staining slightly pink where cut, yellowish pink to brownish in the stipe base, with a faintly acidulous taste and scarcely any smell. Numerous cylindrical cystidia are present on the gills.

Gomphidius roseus lives in mycorrhizal dependence on the pine and spruce, occurring mainly in pine woods on sandy substrates, almost always solitarily. It appears from August to October, most often on woodland paths among grass and moss, and usually in company with *Suillus bovinus,* a mycorrhizal fungus associated with the pine. Not infrequently the close vicinity in which the fruit bodies of both species grow makes their stipes coalesce.

G. roseus is widespread in Europe and Asia, but is rather scarce everywhere, and absent from some regions altogether. It is edible, yet it should rather be protected because of its rarity.

Got Moth Wax Cap

In mycological literature, this *Hygrophorus* has also been described under the name *H. chrysaspis* or *H. melizeus*. It is related to the Ivory Wax Cap. However, while the latter's fruit bodies do not change their white colour, the cap and the gills of the Goat Moth Wax Cap very obviously stain yellow or rust-coloured.

The 3-5 cm. wide cap is convex at first, rounded, with a slightly felted and incurved margin, later expanded, often wavy, thin-fleshed, slimy when moist, and at first whitish. Sooner or later it starts changing colour, to become finally honey-yellow all over. The stipe is 4-8 cm. long, 8-10 cm. thick, cylindrical, attenuated, white at first and then rusty-yellowish, almost entirely granular-pruinose, and minutely scaly only near the apex. The gills are rather thick, broad, deeply decurrent, distant, pallid in youth, and cream-coloured with an ochraceous tinge in age. Towards the cap margin they acquire a rich rusty colour, and blacken when dried. The flesh is whitish, straw, or rusty-yellow under the cuticle, and reacts to potash-lye solution by staining yellow. It is odourless and its taste is mild. The spore print is white. The Goat Moth Wax Cap grows in autumn in deciduous forests, mostly under beeches, but rather sporadically. It is edible.

Also the more robust *H. penarius* has its home in beech forests, particularly on calcareous soils. It is white all over; only the cap centre is of a pale ochre buff. It has a dry stipe, often short, thick and obtusely angular, scaly all along its length. It is also edible.

Pluteus Cervinus

Pluteus cervinus is one of the very common red-spored fungi that grow on stumps and decaying wood logs. Its relatively large fruit bodies consist of a light to dark grey or grey-brown cap and a rather thin white.

The cap is 5-12 cm. wide, bell shaped when young and expanded at maturity, radially filamentous glossy, sometimes cracked when dry.

The stipe is 6-12 cm. long and about 1.5 cm. thick, and firm. A distinguishing feature—shared by all pluteus species – are remote gills, whitish at first, but soon becoming pale pink and ultimately flsh reddish as the spores mature.

Their edges are the same colour as the gills suface.

These species can be found from May to October. This fungs is edible, but is of little nutritional value and of inferior quality. Its flesh is very brittle and watery, so that, they damage easily during transport.

Hence fungi belonging to the genus pluteus are only sporadically collected by mushroom pickers. They can be used for flavouring soups and in fried mushroom mixtures.

Tricholoma Portentosum

Two species of the genus *Tricholoma* occupy a special position among mushrooms collected for culinary purposes, namely *T. portentosum* and *T. flavovirens*. Both these species are excellent edible mushroom.

T. portentosum is a fleshy mushrooms with a greyish to grey-black cap streaked with blackish fibres. The cap is 3-10 cm. wide, umbonate, slightly viscid when moist, without scales, and often with an irregularly wavy margin. The stipe and gills are white but with quite an apparent lemon-yellow tinge. The flesh has a pleasant farinaceous smell and taste.

The favourite habitat of this species are coniferous, mostly pine, forests on a sandy substrate. It appears in clumps and never occurs before late autumn (from September to December). It can be prepared for the table in many ways.

T. portentosum may sometimes be mistaken for the edible but far less palatable *T. argyraceum* which has a fibrillose cap and yellowing gills and stipe. Its smell is similar to that of *T. portentosum,* and it also grows in clumps in autumn in coniferous forests, especially in the outskirts of spruce stands.

Very rarely indeed, *T. portentosum* is confused with the inedible *T. virgatum*. The latter has an acutely conical, ash-grey, shiny cap with ingrown radial fibres. The whole fruit body shows no trace of lemon tint – on the contrary, it is predominantly grey. The acrid-tasting flesh of this species is capable of spoiling any mushroom dish to which *T. virgatum* has been added by mistake.

The only truly poisonous grey-coloured species of *Tricholoma* with which all the above-mentioned species might be confused through utter carelessness is *T. tigrinum,* a large fleshy fungus growing on calcareous forest soils in hilly and mountainous regions; its cap is coarsely scaly from the break-up of the cuticle and the stipe stains red where handled.

Gerronema Chrysophyllum

R. Singer devised the independent genus Gerronema to involve some fungi formerly classified as, members of the genera Omphalina or Mycena.

Even in its present delimitation however the genus Gerronema is rather problematical, it can be defined with difficulty only on the basis of microcharacteristics.

The 2-6 cm. wide cap is yellow brown, in dry weather greyish brown, deeply funnel-shaped in the centre. The broad and rather distant gills deeply decurrent are orange or yolk-yellow. The thickness is about 4-5 mm. thick, and the stipe is cartilaginous. The spore print is also yellowish. The flesh has no specific odour or smell.

This species usually develops clusters on strongly putrescent stumps or fallen spruce or fir trunks in mountain forests and virgin forests. It is edible.

This fungus should not be mistaken for Xeromphalina campanella which also grows on the stumps of conifers. It is infact a related species, but belonging to a different genus. It is smaller in every respect. Its cap only 0.5-1 cm. across, the slender stipe is 1-2 mm. thick, rusty strigose at the base.

It is rusty yellow to rusty brown. The gills are yellow and vein like.

Cantharellus Tubaeformis

This chanterelle has a 2-6 cm. wide cap, irregularly deeply funnel-shaped, at maturity with a strongly lobed and scalloped margin, ochre-brown to yellowish smoky-grey, darker when moist, sparsely dark innately scaly, mostly wrinkled-striate. The gill folds are grey, yellowish grey, sometimes yellowish, entirely hoary-pruinose when old. They are running down the stipe like wrinkles. The stipe is 4-6 cm. long, 0.5-1 cm. thick, yellow, greyish above, mostly compressed and rugose. The whole fruit body is hollow (down to the stipe base). The flesh is relatively thin, greyish yellow, odourless, with a mild taste. The spore print is white.

Cantharellus tubaeformis favours damp soils in conifer forests, mainly in spruce and pine stands. Fairly often it grows from rotting stumps overgrowing with moss. It can be found from the end of summer until late autumn, practically until the first frosts. Having a special liking for acid substrates, it is less abundant on calcareous soils. It is an edible, relatively tasty mushroom.

In professional mycological literature and in mushroom-collectors' manuals we often come across the name *C. infundibuliformis,* by which most authors understand

C. tubaeformis. Other authors, however, assign the designation *C. tubaeformis* to a different species identical with *C. lutescens,* having bright orange, lamelliform gill folds and a golden-yellow orange stipe. It is much rarer and occurs in wet conifer forests.

Amanita Spissa (Fr.) Opiz

Amanita spissa it not poisonous, and many people collect is for table use at times when other edible mushrooms are scarce. However, it is a mushroom of inferior quality; moreover, careless and superficial identification in nature can easily lead to confusion with the deadly poisonous Panther Cap – which in fact is frequently the case. It occurs, sometimes in clumps, in summer and autumn, in both deciduous and coniferous forests, especially on lighter sandy soils. It grows even in long-lasting drought. Its cap is 5-10 cm. (sometimes as much as 20 cm.) wide, smooth and ungrooved at the margin, entirely grey or ashy brown-grey; lighter-coloured and sometimes whitish in wet weather. When dry it is shiny, scattered with many scab-like, whitish or pale grey remnants of the universal veil, often arranged into incomplete concentric rings. The clavate white stipe swells at the base into a conical or turnip-shaped bulb, ending in a hooked and attenuated to pointed excrescence. It has no volva but one or two scarcely discernible rings of wartlike projections. The ring is large, flaring, membranous, distinctly longitudinally grooved, white. The gills are free, broad, crowded and white. The flesh is also

white, occasionally greyish under the cap cuticle, smelling of raw potatoes, with a mild taste. The spore print is white.

Both *A. spissa* and *A. rubescens* are distinguished from the poisonous Panther Cap by the colour of the cap and by a distinctly grooved ring; in *A. rubescens* also by the reddening flesh.

Pine Wood Wax Cap

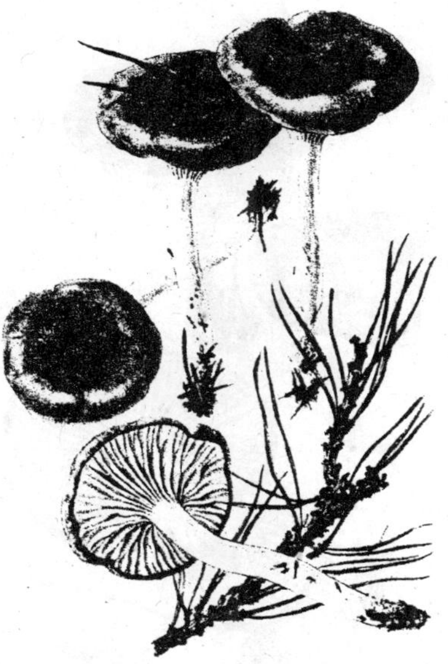

One of the last mushrooms to appear only after the first autumn frosts on the needle carpet of pine forests, is the Pine Wood Wax Cap. It frequently occurs in great profusion under pine trees.

It likes to grow in places exposed to sunlight, in woodland glades and, along forest paths. It is smaller in size. The cap is convex at first then expanded, depressed at the centre later on with an involute margin and is relatively thin fleshed.

The thick cuticle covered with colourless slime, is readily separable its colour is cocoa-brown with a cloudy olive tinge.

It is darkest towards the centre of the cap. The distant relatively thick elastic gills are pale yolk-yellow. The stipe is long and thin cylindrical, coated with colourless slime on a yellowish background with remnants of the viscid veil in the apical part. The flesh has no specific smell and taste. The spore print is white.

The Pine Wood Wax Cap is a relatively tasty edible mushroom. It grows at a time when larger species of edible mushrooms are rapidly disappearing from the wood. It is edible.

Blackening Russula

When visiting a forest in springtime, we may frequently encounter groups of old, blackened, as if mummified fruit bodies which are similar in shape to the genus *Russula*. In fact they usually belong to some species of *Russula* from the group of 'nigrescent' russulas (section *Nigricantes*). Most often it is the Blackening Russula. The robust fruit bodies of this fungus are capable of surviving in the forest for a very long time.

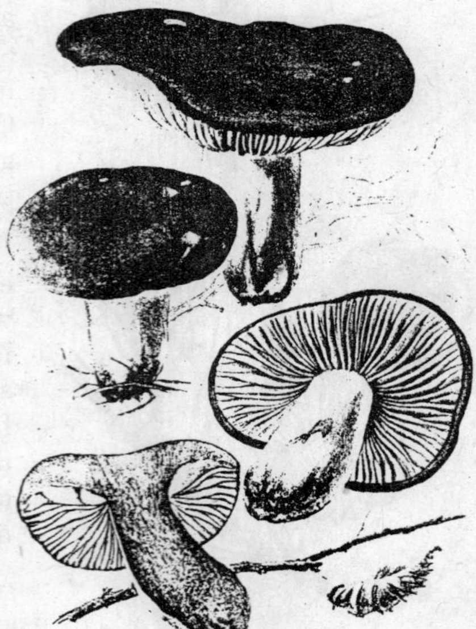

The Blackening Russula occurs in summer and autumn in both deciduous and coniferous forests. Its cap grows to be 8-20 cm. wide; its stipe is 4-10 cm. long and 2-3.5 cm. thick.

The gills are very distant, thick and brittle, and the spore print is pure white. When cut or bruised, the flesh of a fresh fruit body turns brick-red, but later it becomes grey and ultimately black. The flesh of the related *Russula adusta,* found most frequently in pine woods, does not turn red but directly brown to black. The difference between the two species clearly appears also in the gills: these are markedly closer in *R. adusta* than in *R. nigricans.*

Both *R. nigricans* and *R. adusta* are edible mushrooms widely collected in some regions, yet they are not so suitable for eating as it is sometimes maintained. Their flesh is relatively tough, and also their earthy smell is a drawback.

The decayed and blackened fruit bodies of russulas of the section *Nigricantes* sometimes serve as hosts for interesting minute fungi of the genus *Nyctalis*, growing in groups or clusters on the surface of the caps. The two species of *Nyctalis* are fleshy, whitish, at most 2 cm. wide, often bearing incompletely developed gills on their underside. When young, the fruit bodies of *N. lycoperdioides* resemble small puffballs.

Gomphidus Glutinosus

Groups of conspicuously slimy fungi appear in coniferous forests after rains in late summer and in autumn. At first sight they resemble grey-coloured members of the genus suillus coated with a thick layer of gelatinous slime.

Infact they share many of their properties and are moreover closely related to the family, Boletaceae. It has wide cap coated with a readily separable gelatinous layer. At first it is slightly convex, grey or chocolate coloured, then fleshy grey and expanded.

The gills are whitish at first, later they are greyish to blackish, relatively narrow, elastic and, deeply decurrent. The stipe is 4-10 cm. long, and about 1.5 cm. thick. The youngest fruit bodies are covered with a transparent gelatinous veil whose remnants are discrenible in mature fruit bodies in the form of an elevated ridge on the stipe right below the gills. Dry fruit bodies are shiny.

G. glutinosus is edible and very tasty, which also applies to all other Gomphidius species. It is distributed throughout the north temperate zone, wherever spruces or pines have their home.

Caesar's Mushroom, Imperial Mushroom

Caesar's Mushroom is a strictly thermophilous species whose range coincides with that of the warmth-loving flora. In the countries of southern Europe, it is one of the most popular amanitas. In central Europe it occurs very rarely indeed, and is absent from Britain and north-western Europe. It prefers oak and oakl-hornbeam woods on calcareous substrates, yet in southern Europe it occurs even under pines. It grows from July to September.

The wonderfully brilliant-orange to red cap, measuring 8-20 cm. in diameter, is fleshy, smooth and glabrous, or strewn with sporadic white fragments of the universal veil. In youth the universal veil envelops the entire fruit body in the form of a relatively thick white membrane, persisting at maturity as a high, free volva at the base of the stipe. The cap margin is more yellow and finely grooved. The colouring of the cap is variable, forms ranging from white to honey-brown and deep red may occur. Attached to the deep yellow to golden-yellow stipe is a concolorous, grooved, thinly membranous ring. The gills are free, fairly crowded, broad, golden-yellow, covered by a yellow veil in youth. Also the flesh under the cap cuticle and at the apex of the stipe is yellow, sometimes only tinged with yellow to almost white. When submerged in water, it releases a yellow pigment. It has an inconspicuous odour and a pleasant taste. The spore print is yellowish.

Caesar's Mushroom is considered to be one of the most appetizing mushrooms. It was also highly thought of in antiquity, and in Ancient Rome it was regarded as a select delicacy.

Hypholoma Capnoides

Hypholoma capnoides belongs to common *Hypholoma* species, though it is far less common than the Sulphur Tuft and the Brick Red Hypholoma. It forms tufts on stumps and dead roots of some conifers, mainly spruces and pines, mostly in upland forests. Since the fruit bodies develop in the cooler seasons of the year, they may regularly be found both in late autumn and in early spring. *H. capnoides* resembles the Sulphur Tuft in general appearance, size and colouring, but is distinguished by having a differently coloured flesh – it is white or whitish in the stipe as well as in the cap, rusty yellowish only in the stipe base, but never sulphur-yellow. Its taste is not bitter but mild. The flesh is almost odourless or smells slightly of iodoform. In youth the gills are poppyseed grey-blue and later lilac. The stipe is often sinuous, whitish above, otherwise yellowish, rusty-hued in its lower part, finely silky fibrous with white filamentous remnants of the veil. The cap is up to 5 cm. wide, yellowish orange-ochre, darker at the apex, later weathering to paler, with temporary fibres left by the veil in the margin. The spore print is brownish purple-violet. Chrysocystidia are present in the hymenium, which is a common feature of all species of *Hypholoma*.

H. capnoides is edible, some mushroom pickers consider it a good mushroom with a pleasant flavour, especially when pickled in vinegar. Of course, care must be taken not to confuse it with the Sulphur Tuft or the Brick Red Hypholoma.

Shaggy Pholiota

The Shaggy Pholiota is readily recognized by the conspicuous scales covering both cap and stipe. It forms dense tufts on dead wood, mainly on stumps and fallen trunks of both deciduous and coniferous trees.

It is frequently parasitic damaging live tree trunks or roots into which it penetrates through frost cracks or wounds caused by mechanical injury.

It grows mostly in autumn. Some mushroom pickers mistake it for the Honey Fungus whose white spore print represents a substantial difference between the two species.

The cap of the Shaggy Pholiota grows to a size of 12 to 15 cm. in diameter and is coloured in various shades of yellow. The scales covering its surface are rusty brown with upturned or even recurved tips usually arranged in dense circles. The tough stipe also has rusty brown scales on a pale yellowish background.

In young specimens it is connected to the cap, margin by a thick coriaceous veil persisting on the stipe in the form of a woolly tomentose ring.

This gills are adnate to shortly decurrent, fairly crowded, relatively narrow, at first pale olive-yellowish then brownish yellow.

The flesh is yellowish, with a rather pleasant spicy smell, and bitterish taste.

Lyophyllum Fumosum

This mushroom has also been known and described in literature as *Lyophyllum conglobatum,* or still earlier as *Tricholoma conglobatum*. It resembles *L. connatum* in its general appearance, as well as in that its whitish stipes grow out of a common tuberiform base in large numbers. It is distinguished from it by a dark grey, greyish brown to blackish brown cap and whitish gills, which ultimately turn grey. Fruit bodies growing in the dark, e.g. in cellars or other subterranean spaces, are usually white or very light-coloured. The white flesh tastes and smells of cucumber and flour. Quite different is the shape of spores which are subglobose and extremely small.

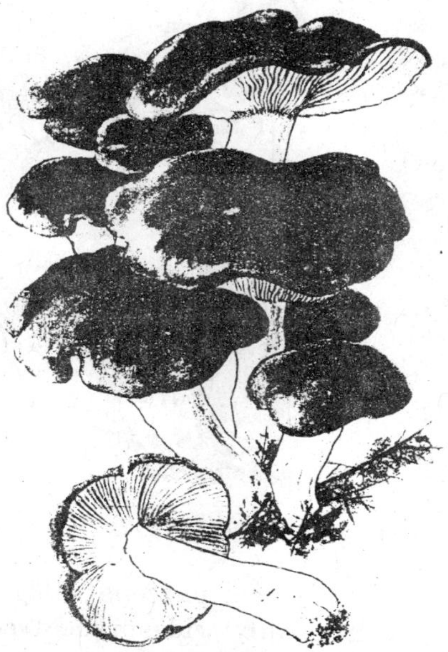

L. fumosum occurs predominantly in places associated within human activity – in yards, along fences, in roadside ditches, rubbish heaps, orchards, gardens, but also in cellars, on piles of excavated earth, or in street pavements. Its occurrence in nature – especially in woods – is rare.

Powerful clumps of *L. fumosum* sometimes contain as many as 100-200 fruit bodies, and the tuberiform structure concealed in the ground, from which the fruit bodies grow, measures approximately 5-10 cm. in diameter. The force developed by growing fruit bodies is remarkable. Reports of pedestrian pavements being uplifted by fruit-body clusters serve to support this fact. *L. fumosum* is an edible mushroom with a pleasing taste.

Honey Fungus, Honey Agaric, Shoestring Mushroom

The generally known Honey Fungus is in fact a complex of several species differing from each other in a number of characteristics of secondary importance, whose taxonomic evaluation has hitherto not been finalized. Therefore the state of knowledge of this species and the distinctions among individual varieties differ from one author to another. As an edible mushroom, the Honey Fungus has been widely collected for culinary purposes for a long time. Yet the ingestion of quite a small part of a raw fruit body causes indigestion. Similar troubles may sometimes arise even after sufficient heat processing. The probability may not be excluded that one of the several closely related species making up this complex is poisonous.

From the point of view of mycological phytopathology, *A. mellea* is a much-feared and generally widespread tree disease, particularly dangerous in its mycelial stage.

Clumps of Honey-Fungus fructifications, more rarely single fruit bodies, grow predominantly on stumps but also on roots and trunks of both living and dead woody plants. The main vegetation period of Honey Fungi is in September and October, but in cooler summers they may appear as early as July. The Honey Fungus is distinguished from caespitose species of the genera *Pholiota* and *Gymnopilus* by the pure white spore print, among other factors.

The related *Amrillaria tabescens* is more plentiful in warmer regions where it grows particularly on oaks.

Agaricus Augustus

Agaricus augustus belongs to the largest Agaricus species. Its cap attains 10-20 cm. in diameter, in youth it is sub-globular, later broadly convex, finally flat. Rusty brown appressed, scales are concentrically arranged on a straw yellow cuticle, the cap centre is brown. The robust stoutly cylindrical, 2-5 cm. thick stipe is slightly enlarged at the base, sometimes swollen into a broad bulb.

It is solid and firm of first, later hollow, bearing a remarkaby large, softly membranous, white superior ring.

The stipe below the ring is covered with large, protruding, wooldy recurved scales, disappearing with age.

The gills are white at first then greyish pink, grey-brown and ultimately chocolate-brown.

The flesh is white, when handled or cut, it turns pale-yellow at first then brown, it smells pleasantly of aniseed and has a mild taste. The spore print is white.

In general it is a rare species, occurring more profusely in some regions only. It is edible.

Glistening Ink Cap

Large clumps of Glistening Ink Caps appear usually after summer rains around the stumps of deciduous trees or

directly attached to them. These clumps consist of a great number of fruit bodies whose caps are 1-4 cm. high, 1-2.5 cm. wide, thin-fleshed, brittle, pale ochre-brown, buff or red-brown. The caps, initially cylindrical or egg-shaped, become bell-shaped or flatly expanded later, up to 4 cm. wide, with an at first grooved, then radially cracking surface. Like the majority of ink caps, mature fruit bodies dissolve into a pulpy matter. The cap surface of young fruit bodies is densely covered with glistening, white or ochre-hued granules; as these are readily deciduous, the caps eventually turn glabrous. The gills are free, crowded, whitish at first, then greyish, mature gills are black and deliquescent. The 4-10 cm. long and 2-5 cm. thick stipe is white, ringless, brittle and hollow, white pruinose to pubescent at first, then glabrous. The context has no specific odour or taste. The spore print is umber- or date-brown.

The Glistening Ink Cap grows from stumps, and often also from the foot of older live trunks of deciduous trees. Young fruit bodies are edible, but after their ingestion it is necessary to avoid alcohol.

Still other similar ink caps may be found on stumps and tree trunks. Their exact identification requires microscopic study.

Strobilurus Esculentus

The genus *Strobilurus* was devised to contain fungi similar in appearance to those of the genus *Collybia,* whose mycelium grows upon the cones of conifers – mainly of spruces and pines – which have been lying about on the ground for some time. Yet it is not only its biology that characterizes this genus. It differs morphologically from similar fungi of the Tricholomataceae family in having a 'hymeniform cap cuticle', i.e. a cuticle made up of globular cells. In fungi belonging to the genus *Strobilurus,* numerous cystidia of characteristic shape are inserted among the hymenial basidia. They differ from the species of *Marasmius* – which they resemble in appearance – in that the dried and again moistened *Strobilurus* fruit bodies do not revive. Almost all species belonging to the genus *Strobilurus* are edible, since they appear in early spring when mushrooms are scarce, their caps (the stipes being too tough) can adequately be used for flavouring spring soups. Three species occur in Europe.

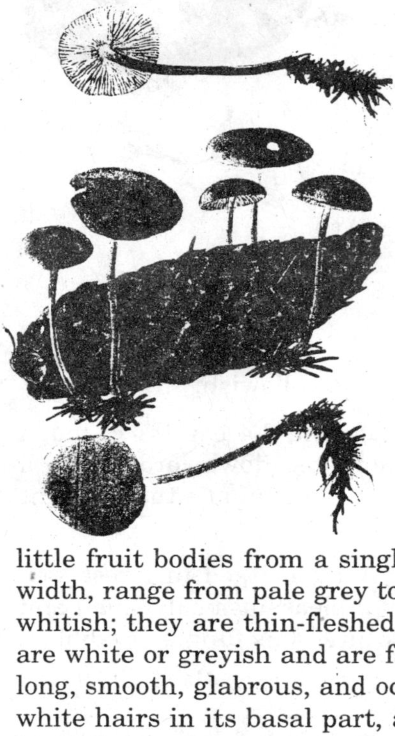

Strobilurus esculentus grows from old spruce cones – usually a larger number of little fruit bodies from a single cone. Its caps, up to 2 cm. in width, range from pale grey to almost chocolate-brown, rarely whitish; they are thin-fleshed, soon flatly expanded; the gills are white or greyish and are fairly crowded. The stipe is thin, long, smooth, glabrous, and ochre-yellowish, coated with long white hairs in its basal part, and if the cone is deeply buried in the humus, this part is proportionally extended to resemble a long string. The other two species, the inedible *S. tenacellus* and the edible *S. stephanocystis,* grow on pine cones.

Catathelasma Imperiale

This is a robust fleshy fungus of the Tricholomataceae group, resembling the genus Tricholoma also in its general appearance.

It is distinguished from the typical representative of the genus Tricholoma by a double annulus on the stipe, as well as by clearly decurrent gills.

The genus catathelesma is microscopically differentiated from the true Tricholoma by anyloid spores and a bilateral gill trama.

It is monotypic genus. The cap is 10-15 cm. broad, at first strongly convex, whitish later slightly convex and broadly depressed, reddish-brown, streaked with brown to molted scaly in the centre.

The gills are dirty white to pale-yellowish.

The stipe is up to 15 cm. long and 3-4 cm. thick, tough fleshed, pale ochre, usually attenuated downwards, with a double membranous, flaring whitish annulus situated right below the gills.

As a rule the stipe is deeply submerged in forest soil. The flesh is white, very tough of a pleasant mealy smell and a mild taste. The spore print is white. This fungus is edible.

Melanoleuca Brevipes

Fungi of the genus *Melanoleuca* are in fact *Tricholoma* species with amyloid, minutely verrucose, finely ornamented spores and characteristic cheilocystidia. They are, however, distinguishable at first sight even without a microscope by their slender, upright stipe with a fibrous surface, and a very regularly circular, relatively thin-fleshed cap that gives the impression of having been turned on a lathe; the cap may be grey, brown, or less often ochre or white. The cap often bears an obtuse umbo at the centre, the gills are always crowded

and regular, white or yellowish. Individual species, about fifty of which have been described, are extremely similar and can be distinguished only with difficulty.

Melanoleuca brevipes occasionally makes its appearance in grassy places beside field paths, in meadows or in gardens. Its cap measures 3-6 cm. in diameter; it is greyish-brown, umber or porphyritic-brown, innately radially fibrous, and depressed-umbonate in the centre. The gills are persistently white, the stipe is concolorous with the cap or somewhat paler, claviform, 2-4 cm. long, 4-8 mm. thick. The flesh is whitish in the cap, brownish in the basal part of the stipe, tasteless and odourless. The spore print is white.

Noteworthy among the other species, edible just like *M. brevipes,* is *M. cognata* with its brownish cap measuring up to 10 cm. in diameter and bright ochre-yellowish gills. It occurs from April to May, and then again in autumn, in both coniferous and mixed forests at path edges, but also in the decaying foliage is old gardens, parks, and on compost heaps.

Rooting Shank

From July to September especially in deciduous forests we may come across a conspicuous fungus on a tall slender stipe, whose cap is strongly glutinous in moist weather; moreover, it attracts attention by its prominent radial wrinkles.

The cap is 4-10 cm. in diameter and only rarely a little larger and is light ochre-brown to yellowish-brown in colour.

The gills are white, deeply emarginate, subdecurrent broad, and distant.

The stipe is 5-10 mm. thick, whose total length is about 20 cm. or more.

A great part of it is hidden in the ground and can be inspected only after a very careful extraction of the whole stipe from the soil.

The lowermost part of the stipe, fusiform at first suddenly passes over into a long gradually attentuating rooting structure, which is white and terminates in a white, thin point.

The Rooting Shank grows solitarily or in smaller groups from very decomposed stumps, from putrescent wood logs or roots concealed in the ground; thus the impression often arises that it is growing straight out of the ground. This mushroom is edible.

Shaggy Parasol

The Shaggy Parasol resembles the Parasol Mushroom but never attains its height. The cap is at first globularly closed, then conically bell-shaped, and finally flatly expanded, 10-15 cm. wide. Its cuticle breaks up into large, overlapping scales which are larger and more recurved than in the Parasol Mushroom, pale-brown at first, darker-hued later. Only the umbo in the cap centre remains persistently smooth. The gills are white, turning in old age orange or red on injury; they are broad, crowded, and free. The stipe is 10-20 cm. high, up to 1.5 cm. thick, cylindrical, tuberously enlarged at the base to a thickness of 4.5 cm, white, hollow, bearing a powerful superior double ring which is freely movable and thus can easily be shifted along the stipe. The injured stipe surface becomes at first saffron-yellow, then red, and finally brown. The flesh is white, fibrous, and staining orange at first and then red when injured; its smell and taste are pleasant.

The Shaggy Parasol grows very profusely in woods of all types, especially in warmer and drier situations. It is edible but only the cap is suitable for eating. It is easily distinguished from the Parasol Mushrooms by the remarkable colour change in the cut flesh. In contrast to the Parasol Mushroom, its stipe surface is not striated. A very light-coloured variety of *M. rhacodes,* var. *hortensis,* may be found growing on compost heaps and manured garden soil. Just like in the Parasol Mushroom, the mycelium of the Shaggy Parasol tends to remain in the same spot for a number of years. Thus it is possible repeatedly to gather the Shaggy Parasol in regular places.

Agaricus Brunnescens

This agaric is better known as *A. bisporus.* It is medium-sized, with a cap rarely exceeding 10 cm. in width, firm in texture and thick-fleshed, globular at first, then flatly bell-shaped, with a central navel-like depression, expanded in older specimens. The cap cuticle is dingy brownish to deep dark-brown, smooth in the centre, with ingrown radially arranged brown fibres near the margin, eventually cracking into appressed brown scales. The cap's margin usually bears soft, wool-like veil remnants. The originally white gills soon turn a vivid pink, then reddish chocolate, finally almost black with whitish edges. The stipe is cylindrical, 3-6 cm. long, 1-2 cm. thick, glabrous, white. The membranous white veil leaves behind a ring; this, however, is not attached to the apex of the stipe but to its lower part – hence it is detachable in the downward direction. The flesh is white, pale flesh-red where cut, then becoming rusty, without a distinctive odour and taste. Microscopically this species is characterized by bisporous basidia.

It usually occurs on manured soil outside forests, along roadsides, in tree rows, gardens, on composts. Today it has become a habit to produce this agaric artificially – both on a small scale and in large growing establishments. It is nowadays identified with *A. hortensis* which is regarded only as its white-capped variety. It grows in similar habitats and is also bisporous. Both are edible.

Tricholoma Flavovirens

The conspicuously coloured Tricholoma flavovirens occupies the foremost place among the edible species of Tricholoma.

It attracts attention not only through its predominantly green-yellow colouring but also by being one of the last larger-sized edible mushrooms profusely growing in pine woods in late autumn.

The cap measuring up to 10 cm. across is at first bell shaped then flattened.

Slightly umbonate at the centre, entirely greenish-yellow or olive-toned, darker radially fibrous, slimy in damp weather.

The gills are emarginate, attached to the stipe by a tooth, and subcrowded.

The flesh is white, yellowish only under the cuticle of both cap and stipe, unchanging when cut, with a pleasant mealy smell and taste. The spore print is white.

It is a mushroom peculiar to dry pine woods on a sandy substrate, where it occurs gregariously from the end of September until November, practically until the first frosts. In recent years, however, it has almost completely disappeared from some regions. It is edible.

Hohenbuehelia Petaloids

Hohenbuehelia petaloides has fleshy fruit bodies, usually either stipeless or with a lateral stipe; a gelatinous layer of hyphae is present in the cap tissue. Under the microscope it is readily distinguished by fairly thick-walled hymenial cystidia, the so-called 'metuloids'. *H. petaloides* has a 3-10 cm. wide and 3-6 cm. long cap, either tan-coloured or only pale brownish, but it may also be ochre-yellow to brownish yellow. The cap is spatulate or fan-shaped, sometimes passing over into a lateral pallid stipe. The cap margin is even or wavy-lobed, in larger specimens rolled up into a funnel, with a coarse flocculose-felted surface, particularly near the stipe. Young mushrooms are almost white. The gills are deeply decurrent, white or with an ochre tinge. The stipe is 5-12 cm. long and 1-2 cm. thick. The flesh is white, odourless and tasteless. The spore print is white.

H. petaloides grows in summer and in autumn on rotten wood, mainly on stumps, usually in damp mountain forests. Its range extends over the temperate zones of both the northern and the southern hemisphere.

It is extremely difficult to make a distinction between the species discussed here and *H. geogenia*. This is of approximately the same size, but its cap is a darker yellowish brown. It grows on the grassy ground of deciduous and coniferous forests. Both species are edible.

Slimy Beech Tuft

Old beech forests harbour a large number of fungus species; the remarkable Slimy Beech Tuft grows on dying or already dead, but still standing beech trunks. Its dazzling porcelain white fruit bodies, coated with colourless slime in wet weather, sometimes are scattered in large clumps over the moss-overgrown trunks of ancient beech giants in mountain virgin forests. In some regions with a relatively damp climate, the Slimy Beech Tuft occurs at lower elevations as well.

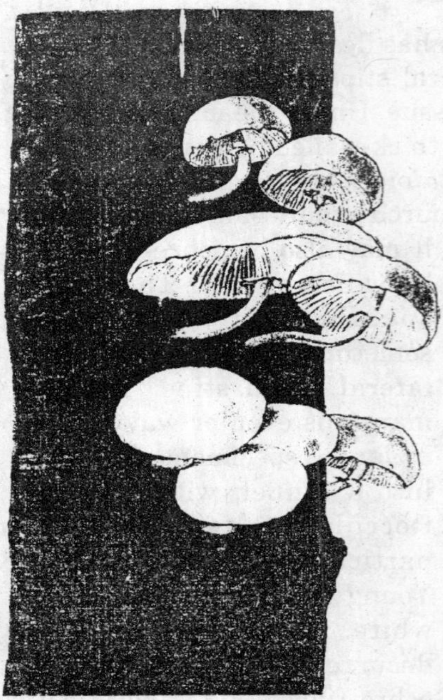

The 3-10 cm. wide cap is broadly bell-shaped, thin-fleshed, and hyaline. The thinly cylindrical stipe is approximately the same length as the diameter of the cap or somewhat shorter, 0.5-1 cm. thick, tough, with a broad, glutinous superior annulus. The gills are subdecurrent, distant, and ventricose. The context has no distinctive smell nor taste. The spore print is pure white.

Large, thin-walled, colourless, lageniform cystidia, 80-90 μm long and 15-30 μm broad, develop on the gills.

The Slimy Beech Tuft grows from July to October. This remarkable fungus is not only aesthetically impressive: it is also of primary importance in biochemistry. Its mycelium produces the antibiotic mucidin, which is effectively applied in treating some skin diseases, especially mycoses. The Slimy Beech Tuft is an edible fungus. Nevertheless, the excessive sliminess of its small fruit bodies and their negligible fleshiness render it unsuitable for table use. On the contrary, it deserves protection, as the rapid decline of its habitats is accompanied by the disappearance of this fungus.

Tricholomopsis Rutilans

The genus Tricholomopsis was separated from Tricholoma on the basis of a combination of the following characteristics; it includes lignicolous, fungi with a central stipe and a yellow ground colour which becomes most apparent in the gills and in the cortex; innate reddish, green, olive, blackish or reddish brown scales, are present not only in the cap cuticle but also on the stipe.

The flesh is rather tough, the spore print is white, the spores are smooth and nonamyloid. Large cheilocystidia are present on the gill edges.

This fungus occasionally grows, out of dead roots or wood logs. It is particularly abundant in spruce and pine forests on acid substrates.

The fruit bodies appear either solitarily or

in smaller clumps. They have a 5-15 cm. wide cap, campaniform at first, soon flatly expanded. The cuticle is purple red over a yellow ground colour.

Young fruit bodies are suitable for eating in mushroom mixtures. Older fruit bodies when cooked smell unpleasantly, of earth or mice, but their taste remains unimpaired.

Shaggy Mane, Shaggy Cap, Lawyer's Wig

The fruit bodies of most ink caps develop at an amazingly rapid rate: some species need only a few hours to proceed from the germ past the ripening of spores to the deliquescence of mature fruit bodies.

On the large-sized and more fleshy ink caps, it is the Shaggy Mane that occurs in greatest abundance. It can be found in manured soil, in garden grass, along roadsides and on the edge of woods. The cap varies considerably in shape; the typical form is at first longly cylindrical but occasionally may be subglobular, 5-10 cm. high, 3-6 cm. wide, later conically convex, white when young, later with an ochre-hued apex, its entire surface being subtomentose-scaly. Mature fruit bodies colour brown from the margin, the overlapping scales are in part concentrically arranged. The caps of mature specimens turn entirely black and dissolve. In the mature fruit body it is gills that are first subjected to autolysis, and because the spore print is dark brown to black, the caps of the Shaggy Mane are transformed, together with the gills, into an ink-black pulpy matter. The cylindrical stipe may be as much as

15 cm. long and 1-1.5 cm. thick, white, with a slightly thickened base, tubularly hollow, with a narrow, free, almost movable ring which soon falls off and sometimes persists at the stipe base. The crowded, remote gills are at first white, then rapidly turning pink from the cap margin, later they become red-brown and finally black. Young fruit bodies are edible and tasty, but it is not recommended to accompany any meals containing them with alcoholic drinks.

Velvet Shank, Winter Fungus

Fresh fungus fruit bodies can be gathered even during a mild winter. The Velvet Shank is one of the few species that grow at low temperatures and are capable of tolerating a fall to below freezing point. Its clumps can often be seen growing on live trees in insufficiently healed scars left by detached branches and on stumps. They also grow on dead trunks of various deciduous trees, but rarely on conifers.

The cap attracts attention by its ochre-yellow colouring, almost chestnut-brown at the apex. Ecceptionally it can attain as much as 10 cm. in diameter, but its usual width is 2-5 cm. It is expanded-convex, with thin and elastic flesh, sticky in wet weather. The gills are emarginate, crowded, white, then yellowish. The stipe, narrowing downwards, is sometimes tapering to a rootlike extension; it is yellow above, black-brown below, entirely covered with velvety tomentum. The spore print is white. The tasteless flesh smells faintly of raw fish meat.

It is only the caps of the Velvet Shank that are gathered, the stipes being much too tough. Recent research has revealed the extract of raw fruit bodies to be poisonous. Cooked fruit bodies, however, can be eaten. The Velvet Shank is particularly well-known to mushroom enthusiasts living in large cities: they see it in winter in parks and gardens.

Clumps of the poisonous Sulphur Tuft (*Hypholoma fasciculare*) very commonly appear on stumps. Its gills are yellowish green, the same colour as the flesh which is, moreover, intensely bitter. Hence only a great deal of carelessness or ignorance might bring about the confusion of the two species.

Krombholziella Oxydabilis

This mushroom consists a dark brown cap, sometimes grey-brown, with a blakish tinge, occasionally ochre to pale brown.

The stipe is grey to blackish scaly on a white background. The originally white flesh stains pink to red when cut.

The reddened places never turn violet or black. The flesh in the base of the stipe is often of a dingy pale yellow or only yellow.

The cap cuticle is slightly dark fibrillose, almost cream coloured, but not felted. It is dry, turning viscid only in long lasting rains but soon drying again.

It is common particularly in northern and northeastern Europe.

The flesh of *K. Oxydabilis* sometime found in bruises red.

K. Thalassina has an ash-green cap. The bottom part of its stipe is green, it also grows under birches.

All these species are edible and tasty.

Marasmius Scorodonius

Many fungus species can be identified by a specific persistent smell or aroma that typifies the species in question. This attribute is consequently of paramount importance in identification, and a keen sense of smell is invaluable to any mushroom collector. The odour of garlic is characteristic of *Marasmius scorodonius;* it becomes apparent after rubbing dry fruit bodies between the fingers, but is of course most intense in fresh fungi. This odour has inspired some mushroom lovers into adding the species to their food as a substitute for garlic; unfortunately the aroma almost entirely disappears is cooking.

M. scorodonius is most commonly found in sparse coniferous forests, in their outskirts and in heather moors – that is, mainly on acid soils. It grows on dead twigs, roots, but also on needles and stumps. The cap is 1-2 cm. across (rarely larger), thinly leathery, light flesh-brown or buff pinkish to off-white, often sulcate-striate, flatly expanded. When desiccated, it can be revived by moistening. The tough and elastic, only 1-2 mm thick stipe is completely glabrous, shiny red-brown. The gills are quite dense, prue white, adnate. The spore print is white.

In similar habitats, *M. androsaceus* and *Micromphale perforans* grow out of the decaying needle carpet. The former is smaller and odourless, similar in colour to *M. scorodonius* except for its black and glossy, horsehair-like stipe. *M. perforans* is approximately the same size but has a pubescent stipe and an offensive odour of old rotting cabbage.

Greasy Tough Shank

The Greasy Tough Shank has a 4-8 cm. wide cap, bell-shaped when young, later expanded, sometimes slightly and obtusely umbonate in the centre, deep red-brown, chestnut-brown in wet weather, pale cocoa-brown when dry. The cap cuticle is glabrous and smooth. The stipe is 4-8 cm. long, 5-20 mm. thick, cylindrical, markedly clavately enlarged towards the base, soft, coarsely longitudinally fibrous and sulcate. The apical part of the stipe below the gills is lighter to whitish, the rest is red-brown. The gills are deeply emarginate and adnexed, crowded, rather broad, whitish. The flesh is white, sometimes faintly smelling of rancid butter, with a mild taste. The spore print is pure white. The Greasy Tough Shank is a common species inhabiting deciduous, mixed and coniferous forests. It grows in leaves and needles, usually in groups, from July to November, practically till the first frosts. It is edible; only caps are gathered.

Collybia asema is a closely related species considered by some mycologists to be a variety of the Greasy Tough Shank.

Its similarity to the latter in both size and appearance is striking, the only difference being that *C. asema* has an ash-grey or greenish grey cap with an almost dark brown central umbo. The strongly hygrophanous cap fades to a whitish shade in dry weather. Also the stipe is light grey or brownish grey. *C. asema* grows mainly in coniferous and mixed forests, and, in some regions, it is more common than the Greasy Tough Shank. It is also edible.

Tricholoma Caligatum

This fungus is much better known in Japan than in Europe, being the most popular edible mushroom there, where it is called *matsu-take,* i.e. 'pine mushroom'. In Japanese mycological literature it is most frequently referred to as *Armillaria matsu-take.* In Japan it mostly grows in *Pinus densiflora* woods, less frequently under spruces or firs, rarely also under *Tsuga sieboldii.* It is highly esteemed for its taste and smell. It grows from October to November, yet in a rainy spring it may occur as early as May and June. Pine-mushroom hunting is a Japanese national sport, particularly for city dwellers with pine woods in their neighbourhood.

T. caligatum is widespread throughout the north temperate zone, yet in Europe it is extremely rare. It occurs more abundantly in warmer regions – the more northwards, the more sporadic is its occurrence. Robust, thick- and firm-fleshed fruit bodies usually grow solitarily. The greater part of the 10-15 cm. wide cap is covered with large, appressed, fibrous scales, rusty to sepia-brown, the whitish ground colour being visible among the scales. A thick veil joins the cap margin to the stipe where it leaves a broad, membranous annulus that later withers. The gills are crowded, white, then cream, and brown-spotted at the edge. The flesh is white or cream-coloured, with a sweetish or fruity smell and a mild taste.

Common Field Agaric, Meadow Mushroom

***Agaricus campestris* L. ex. Fr.**

The Common Field Agaric is the best known and most commonly collected of all the agarics growing in the wild. Its cap is 5-12 cm. wide, at first globularly closed, then semi-globular, finally flatly expanded with a silky-fibrous surface, sometimes cracking into scales, white, in old fruit bodies tinged slightly with yellow to brown after handling. The gills remain pink for a long time, gradually discolouring to red; when old they range in shades from chocolate to black-brown. The stipe is 5-10 cm. long, stoutly cylindrical, 2-4 cm. thick, white or slightly pinkish and rusty or brownish in the lower part, tapering to a point at the base. The membranous ring is white and readily detachable. The flesh is white, either turning pale pink or remaining unchanged on injury, with a pleasant, somewhat fruity smell and taste.

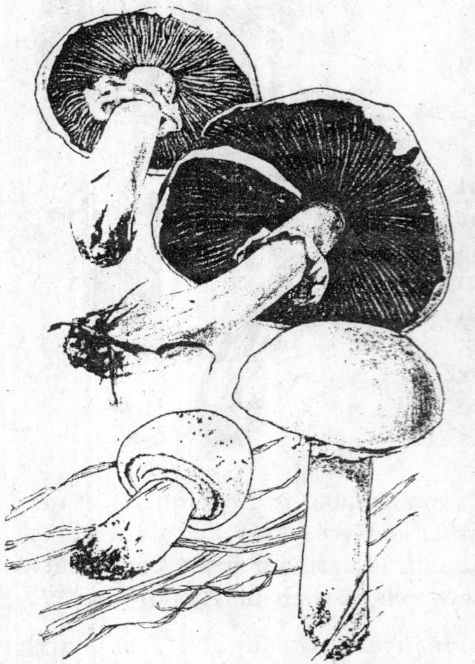

The Common Field Agaric occurs outside the woodlands in manured soils, e.g. in meadows, fields, ditches and compost heaps. It grows in summer and autumn; large numbers of fruit bodies usually emerge after heavy rainfalls. Due to its strong variability within the species, it forms numerous varieties.

The similar Horse Agaric (*A. arvensis*), in spite of the fact that its Latin specific name means 'of the field', is a forest species favouring the forest humus of drier woods and shrub growths along forest edges. It is distinguished mainly by the cap and stipe cuticle rapidly becoming bright yellow on injury, as well as by an intense and pleasant scent rather like aniseed. When young the Horse Agaric has only pale flesh-coloured to grey-reddish gills.

9

Collection and Identification of Fungi

Some edible mushrooms may be gathered from spring till autumn on tree stumps or trunks, e.g. Kuehneromyces mutabilis.

In the period of the most intensive vegetation of fungi, we may find mushrooms almost in every forest, but not all forest growths are equally rich in mushrooms.

There are mushrooms pickers who mark the trees in whose vicinity they have found, for example various boleti.

Such a place often remains fertile for a number of years, since these mushrooms live in mycorrhizal dependence on a particular woody plant and regularly form fruit bodies at a certain time. There is a difference between those who gather mushrooms exclusively for practical purposes, that is for enriching their diet, and those who regard fungi as an object of scientific interest. However, both kinds of mushrooming can be pursued as a hobby.

The practical mushroom pickers' attention is usually focused on a limited number of fungus species which he can safely recognize and which he prefers. Only if he fails to find his favourite species, or on the recommendation of other collectors will he also take other edible mushrooms.

The mushroom picker's, actual season is relatively short.

In European conditions it is concentrated at the end of summer and the beginning of autumn, that is from August to October, when mushrooms grow most abundantly.

It is also necessary to take account of the weather, as the most favourable conditions for mushrooming arrive after heavy rainfalls followed by the warm weather.

Mushrooms like neither wind nor excessive temperature variations.

However edible mushrooms may be found even in spring, if the weather is both moist and warm.

This is the time when large-sized spring Discomycetes appears – morels and members of the genera Discina and Gyromitra, of the gill fungi it is Hygrophorus marzuolus, strobilurus species growing on cones, later on St. Georges' Mushroom and Entoloma clypeatum. Mushroom pickers prefer sunny places sheltered from wind, such as southward-facing woodland margins, lawns, gardens, forest clearings, etc.

It is advisable to look for light deciduous or mixed forests, while in extensive, monotonous conifer forests, no mushrooms are usually found in spring time. In summer, if the weather is exceedingly dry, situations on shady slopes, facing northwards or in river and brook valleys should be looked for.

A knowledge of woody plants helps in the search for mycorrhizal mushrooms (a number of boleti grow exclusively under aspens, birches and poplars). Wherever the countryside is richer in its natural aspect, particularly as concerns the structure and specific variety of woody plants, its mycoflora also displays a greater abundance of species. Geophilous fungi growing in the humus, fallen foliage or needles prefer forests without any herbal or shrubby undergrowth. Where the wood is excessively overgrown with grass and a buoyant, dense herbal vegetation, or where the soil is covered with a continuous layer of branches left behind after wood sifting or wood production, geophilous fungi are usually scarce.

Edible mushrooms are also gathered outside the wood, e.g. in meadows, fields and pastures. Whole swarms of champignons (*Agaricus*) often appear there after summer rainfalls. After the height of the season, which is usually in September, several lignicolous species may be found growing on stumps and decaying tree trunks in the cool season – perhaps also during a mild winter. This applies to the Winter Fungus or Velvet Shank (*Flammulina velutipes*) and the Oyster Fungus (*Pleurotus ostreatus*).

Anybody whose concern for fungi is primarily based on his scientific interest can pursue his hobby on a substantially larger scale: he may collect and study fungi almost all the year round, depending only on the group he selects. Leaving aside all the microscopic fungi wherein scarcely any amateur mycologist takes interest nowadays, there still remains a number of interesting and often little-known species (sometimes even unknown ones, hitherto nor described) whose seeking in nature, microscopic examination and identification yield much satisfaction to anybody whose attention they have aroused by their various shapes and way of life. They can be collected from spring to winter in most various habitats within and outside the forest. During our excursion we proceed as slowly as possible, focussing our attention both on the ground and on stumps, dead branches either scattered about or jutting out, trunks of uprooted trees, but also on the remnants of herbs and grasses, decaying leaves, needles, animal excrements, burned-over areas, roadside edges overgrown with low moss, and ditch banks or old walls. Good results follow from a systematic inspection of some selected small ecotypes (e.g. areas around springs, alder coppices), examined in regular intervals throughout the year or for several years, while all findings are recorded and documented and the major part of the higher fungi consists of so-called pileate fungi, whose fruit bodies are usually differentiated into a stipe and a cap. The pileate fungi moreover include fruit bodies of stomach fungi (Gasteromycetes) – puffballs (*Lycoperdon*), members of the genera *Bovista* and *Scleroderma,* but also other species whose fruit bodies are conspicuous in shape, e.g. stinkhorns (*Phallus*) and earth stars (*Geastrum*). Some species of larger cup fungi (Discomycetes), flask fungi (Pyrenomycetes) and non-gilled fungi (Aphyllophorales) belong here as well. The fungi included here predominantly belong to the Basidiomycetes, while the representation of the Ascomycetes is relatively small.

All the fungi whose fruit bodies are collected in the open air for the purposes of study and identification should immediately be deposited with care, one by one, in metal or plastic boxes of various size. Non-gilled lignicolous fungi are cut off together with a part of the substrate and wrapped into a piece of newspaper or put into a paper bag. We must not forget to put down the name of the host or, rather better, to enclose a fragment of the plant whereon the fungus has been found; in lignicolous fungi either a fresh or a dry leaf of the woody plant in question will do. It has proved useful to add to the fresh small and fragile gill-fungi fruit bodies a fresh leaf of any green plant which, by evaporating water, keeps the air in the box humid and thus prevents the fungus from desiccating. After our return from the expedition, we put down on the margin of the paper in which the separate finds are wrapped the data concerning the locality, habitat or substrate, and the date of collection. After processing the material the notes are clipped out and deposited together with the collected specimens first into the drier and then into a herbarium envelope. They are crucial for writing the herbarium label. In lignicolous fungi, particularly in those having resupinate fruit bodies, the quickest possible desiccation of the collected material is necessary: only in this way is it possible perfectly to preserve on the exsiccations the qualities of the hymenium (especially the shape of basidia) which are of primary importance for the identification.

The present classification of gill fungi is predominantly based on microcharacteristics observable only under the microscope (shape, size, ornamentation of spores, fruit-body tissue, structure of the gill trama, cuticle of the cap and stipe, etc.). Notwithstanding, it is of extraordinary importance to work out a detailed description, or at least notes, concerning all the perceivable characteristics involved (colour, taste, smell, chemical reaction of the context); these characteristics usually disappear in the desiccated fruit body, or change so much that they can no more be reliably ascertained. The spores and the fruit-body structure can of course be most easily studied on fresh fruit bodies, yet they can also be examined on well dried exsiccations at any later time. Fruit

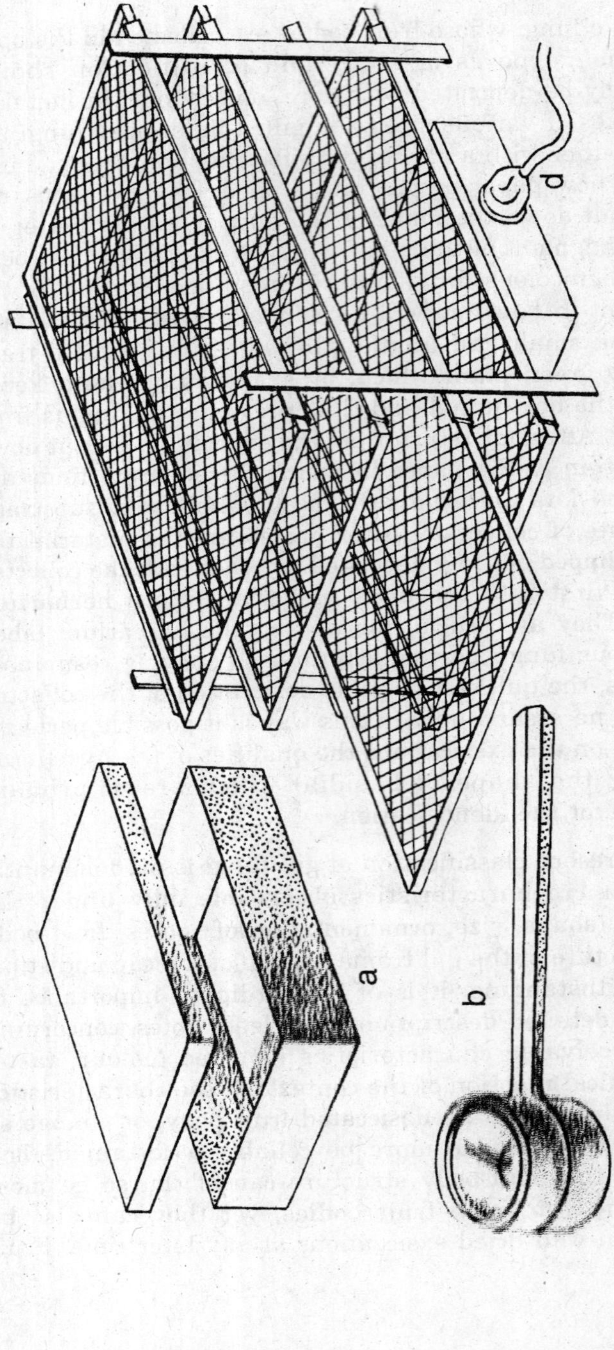

Some aids for the collection and preparation of fungi: (*a*) plexiglass box; (*b*) magnifying glass; (*c*) knife; (*d*) electric drier.

bodies are usually dried as a whole, only large specimens are cut lengthwise into two or more parts. They are dried in moderate (40-60ºC) but permanent temperatures without interruption, until they become bone-hard or fragile. Then they are placed for a few hours into a plastic bag containing a piece of paper or cotton wool dipped in water. The fruit bodies absorb a small amount of moisture, which makes them pliable. Only then do we put them into herbarium envelopes of tough, strong paper, and slightly press them by hand. This will prevent the fruit body from breaking when further handled. The collection of exsiccati must be kept dry and disinfected at least once a year (carbon-disulphide fumes being the best disinfectant), and protected against insects by chemical agents.

An important distinguishing feature of fungi is the colour of the spore print. The spore print is obtained by cutting off the cap of a fresh, ripe fruit body from the stipe and placing it flat, gills downwards, on a piece of white paper (eventually on a black one, if a white spore print is expected). The cap together with the paper are either inserted into a plastic bag, enclosed in a tin or covered with a bowl. As a rule, the spore print appears on the paper in a few hours. When using a microscope, we jot down and draw the shape and size of spores (these are measured, just like other micro-elements, by an ocular micrometer), as well as the ornamentation of the spore wall, ascertained with the aid of an oil immersion lens after applying Melzer's reagent or cotton blue to obtain the required colour reaction. Further, sections are cut through the gills and the cuticle of both the cap and the stipe. Minute parts of tissue from the fruit-body surface are cut off with a razor blade and turned into so-called crush mounts; these are far less elaborate and usually prove sufficient for identification. A tiny piece of tissue, perhaps a fraction of a gill, is mounted in a droplet of water on a microscope slide, and a small glass cover slip is placed over and slightly pressed. Microscopic mounts of fresh fruit bodies are prepared in water, those of exsiccations in an aquaeous ammonia solution, in a 10-per cent potassium hydroxide (KOH) or lactophenol.

Every observation is recorded and sketched on a separate sheet of paper. On principle, each finding and herbarium item is kept separately. This method has turned out to be the best: it allows an easy manipulation and a large number of descriptions can easily be classified according to a certain system, or alphabetically, and can quickly be looked up again. Descriptions may be complemented with coloured sketches or photographs. Everybody who takes a serious interest in mycology must become acquainted with the fundamentals of working with a microscope, for nowadays it has become absolutely impossible to do without a good microscope in pursuing this branch of study. Anybody who can draw well is on the winning side, because no microphotograph, and frequently not even a macrophotograph, good as it may be, can fully replace an exact and readily made sketch of the fruit bodies under examination.

Today the identification of fungi is aided by an extensive literature – from general keys up to monographic elaborations of genera or whole families. It is advisable to use the most modern versions and continuously to compare them with older works. The best solution for a beginner is to contact a specialist in the respective fungus group who will help him to overcome the first and most difficult obstacles.

10

The Morphology of Inedible Fungi

The category of inedible fungi is one, which is most problematic and most difficult to define.

For the most part this designation covers fungi which are neither edible, nor poisonous, or whose edibility is still open to doubt due to lack of exact information. As a rule they possess qualities making them unsuitable for table use; too small fruit bodies unpleasant smell and taste excessively tough flesh, and sometimes they inspire distate by their appearance or place of occurrence. Most of these features, however, can be purely subjective with the spreading of knowledge of individual fungus, species some of them are gradually being reclassed from the category of inedible fungi into that of poisonous fungi.

In future it might be expedient to omit the category of inedible fungi altogether, and to distinguish only mushrooms suitable for table use and safely verified as regards their harmlessness to health, and fungi detrimental to the human organism—that is poisonous.

Entoloma Hirtipes

Entoloma Hirtipes is a rare species sometimes also classified under the generic name Nolanea. Its cap is 3-5 cm. wide rarely larger, campanulate conical, grey-brown with gills slightly showing through in wet weather. The stipe is 4-6 mm. thick with a white tomentose base swollen to a thickness of about 1 cm. the same colour as the cap very brittle.

The gills are ventricose, broad, at first whitish and finally grey. The flesh, particularly after handling, emits an intensive and penetrating smell of codliver oil, the spore print is pink.

From April onwards, *E. hirtipes* grows on the ground in grass mostly in coniferous forests.

It is inedible. The fungus is closely related to *E. mammosum* which grows in summer and in autumn and also prefers grassy habitates, but more often outside the forest.

E. mammosum is smaller, its cap is usually umbonate sometimes with a sharp point. Similar grassy places are inhabited in spring by the poisonous Nolanea verna. This fungus, however becomes very dark in wet weather.

All these pink-spored or red-spored gill fungi are distinguished with difficulty. Since they can so easily be confused, the practical mushroom picker had better avoid them altogether.

Wood Woolly-foot

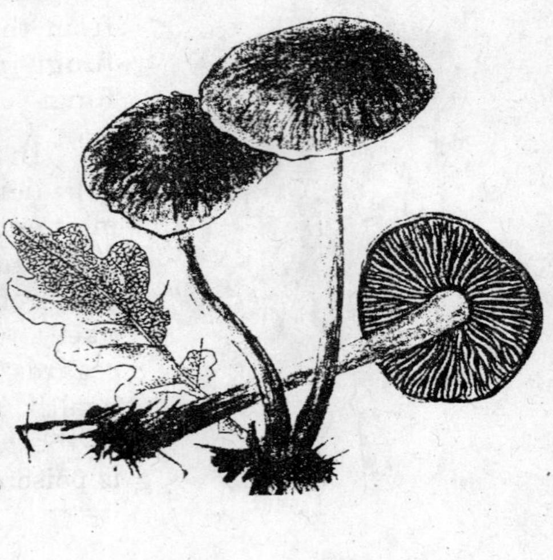

Not all species in the relatively large genus *Collybia* resemble the widely known Russet Shank (*Collybia dryophila*). The Wood Woolly-foot is dissimilar both in colour and—particularly—in its markedly distant gills. The 2-5 cm. wide cap is planoconvex, later expanded, with a slightly furrowed margin and thin, elastic context, light flesh-brown when moist, pale buff when dry, glabrous. The gills are at first pale yellow to yellow-brown, then pale brown. The stipe is relatively long, thinly cylindrical, 3-5 mm. thick, slightly inflated at the base, strigose-felted, yellowish to reddish brown, elastic. The flesh is pale yellow, with a burning acrid taste and an acidulous smell. The spore print is white.

The Wood Woolly-foot occurs profusely throughout the summer and early autumn. It grows either singly or in sparse clusters in rotting leaves heaped up in higher layers in deciduous and mixed forests; occasionally it can be found also in conifer forests. Its acrid taste makes it unsuitable for eating.

Not uncommon in similar habitats is *C. confluens*. It often forms dense clusters consisting of many dozens of specimens. This species has a pale tan or brownish, sometimes whitish, cap, and very crowded, narrow, whitish, later brownish gills. The contxt is not acrid. The stipe base is densely overgrown with a whitish mycelium enveloping the leaves, needles or twigs from which the fruit bodies grow. *C. confluens* is inedible.

Mycena Fagetorum

Mycena fagetorum is a rare species whose fruit bodies do not appear before late autumn, in October and November, in the rotting foliage covering the floor of beech woods. Its specific name was derived from its association with the beech, which also determines the type of its habitat. It may grow solitarily but usually is found in larger communities. The 1-3 cm. wide cap is at

first semiglobular or conical, then very slightly convex, sometimes almost flat, in wet weather distinctly striate, pallid or whitish, pale grey-brown round the centre and in the striate parts, with a frequently pellucid and colourless central part, whitish all over when dry. The gills are distant, whitish or slightly grey-brown, adnate to shortly decurrent, and relatively elastic. The stipe is 3-8 cm. long, 1.5-2.5 cm. thick, dark or olive-grey, later grey-brown, hyaline, hollow, attached to the leaves by its white, hairy base. The spore print is white. The cheilocystidia have branched excrescences at the top. The flesh has a farinaceous smell and a mild taste.

Like the majority of *Mycena* species, *M. fagetorum* is inedible. In spite of their profuse occurrence, members of the genus *Mycena* are of no importance for practical mushroom pickers because of their small size. Many of them, however, occur in such great numbers that they become conspicuous towards the end of summer and in autumn.

Gymnopilus Spectabilis

The genus Gymnopilus includes several larger-sized and relatively flesh ochre golden yellow to oragne-yellow fungi. They all have a somewhat slimy cap when moist, a cuticle turning black in lyes, a rusty brown spore print verrucose spores, and cystidia turning green when touched with cotton blue.

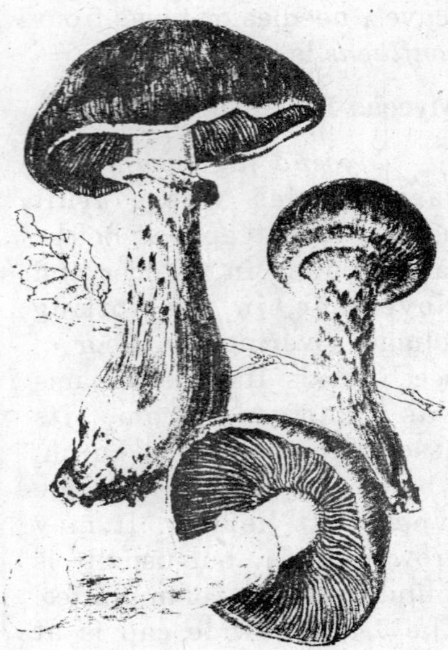

They are for the most part, lignicolous fungi. *G. spectabilis* is one of the sturdiest among them; its cap is as much as 10 cm. wide, exceptionally even larger, it is almost, velvety felted, neither shiny nor viscid, semiglobularly convex, later expanded, with no umbo or only lightly and obtusely umbonate, later cracking into patches, with a slight purple flush here and there. The stipe is pale yellow, cylindrical, often clavately thickened towards the base, turminating in a roat like extension, up to 20 cm. long.

A membranous, yellowish and persistent ring is located in its apical part, and the stipe above the ring is whitish.

The gills are pale yellow brown, later almost golden-yellow, brownish, crowded, narrow and slightly decurrent.

It is inedible.

Lentinus Lepideus

Lentinus lepideus is one of the commonest fungi of the northern hemisphere growing on worked wood: beams, piles and railway sleepers. It prefers to grow on spruce, pine, fir and larch wood. The 5-15 cm. wide cap is at first convex, then expanded, depressed at the centre, in youth with an involute and matted-tomentose margin, with thick and elastic flesh, bestrewn with large, at first ochraceous, later brown scales. The stipe is up to 7 cm. long, 1-1.5 cm. thick, solid, tough and elastic, concolorous with the cap, usually attenuated downwards into a blackish rootlike extension. It is covered with a large number of recurved, brownish, fibrous scales but often grows glabrous. The stipe apex is glabrous and smooth. The annulus is scaly, retrocurved, sulcate on the outside; usually it is only indicated. The gills are broadly adnate, subdecurrent, broad and distant, white at first, then straw-yellowish, almost lemon-yellow near the cap margin, with irregularly dentate edges. The flesh is white and relatively soft in youth, but firm and coriaceous when old, as hard as bone when dry; it has a pleasant fungussy or aniseed smell and an inconspicuous taste. The spore print is white.

L. lepideus is a considerably versiform fungus. The caps of fruit bodies growing in darkness (e.g. in mines) is either underdeveloped or absent altogether—thus they form only an abnormally elongated, often deformed stipe branching like antlers. *L. lepideus* is inedible.

Tubaria Conspersa

Small, inconspicuous and mostly uniformly cinnamon-brown fungi, attracting attention by their occurrence in clusters, make their appearance in summer and in autumn on bare, moist ground shaded by trees or shrubs, sometimes also on decaying plant remains. They usually belong to some species of the genus *Tubaria*. Only a closer acquaintance with their microcharacteristics makes their exact identification possible.

Tubaria conspersa belongs to the most abundant members of this group. The deep cinnamon-brown ground colour of the cap cuticle is covered in a fine greyish white felt, splitting into a fibrous-scaly pattern later on; in very young fruit bodies the cap margin is connected with the stipe by a cobwebby veil. The cap is only 8-25 mm. wide, almost semiglobular in youth, then slightly and roundly convex, somewhat depressed in the centre, with a sparsely grooved margin, thin-fleshed throughout. The stipe is 2-3.5 cm. long, 1.5-3 mm. thick, paler than the cap, enveloped at the base by a white felted mycelium. The gills are broadly adnate and shortly decurrent, broad, distant, deep cinnamon-brown. The flesh is odourless and has a mild taste. The spore print is brown. The cheilocystidia are cylindrical to clavate, often irregularly sinuous. *T. conspersa* is inedible.

During a mild winter and in early spring, very similar *Tubaria* species—*T. pellucida* and *T. furfuracea*—may often be found in grassy patches in fields and along roadsides, outside the wood as a rule. Yet these species cannot be safely distinguished from each other without a microscope.

Little Wheel Toadstool

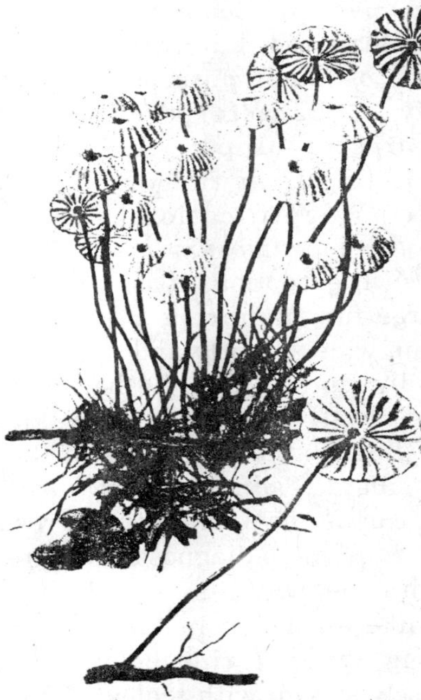

The caps of the Little Wheel Toadstool look like miniature filigree mud-pies, flooding whole areas in their thousands and enlivening the forest twilight by their dazzling whiteness.

It prefers damp and dark localities along forest brooks, mainly in growths. In summer and in autumn it occupies a great many fallen twigs, needles, stumps, and sometimes also decaying leaves.

This species is noteworthy for the form striation and colour of the cap contrasting with the bristly dark stipe.

The gills are coalescing into a collarium along the circumference of the stipe this attribute serves to distinguish the Little Wheel Toadstool with no difficulty. This fungus turns dark with age and fades in dry conditions.

These fungi may be hunted for in lawns, gardens, pastures and unmown meadows after summer rains.

Hebeloma Radicosum

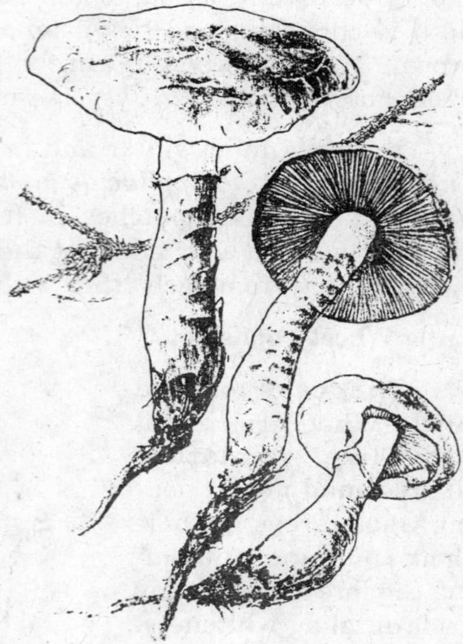

Though fungi of the genus *Hebeloma* can easily be recognized at first glance, their assignment to individual species implies a perplexing mental challenge even for an expert. A great many species have been described, but only a few of them possess truly characteristic features or properties facilitating their immediate identification. *Hebeloma radicosum* is one such exception. It is a large fungus having a 12 cm. wide, slightly and broadly convex, firm-fleshed cap. The cap cuticle is sticky when moist, smooth when dry, glabrous, pale earth-coloured, with fine fibrous remnants of the white veil at the margin. The gills are adnate, emarginate with tooth, pallid at first, then earth-brown, almost red-brown as the spores mature. Its most important diagnostic feature is a stout, firm-fleshed to tough stipe, with a swollen base rapidly tapering into a long root-like extension in the soil. At the stem apex there is a flaring superior ring below which the whitish surface is sparsely strewn with scales. The flesh is whitish, rusty brown in the stipe, and pleasantly smelling of bitter almonds. The spore print is a pale earth-brown.

Usually, *H. radicosum* grows solitarily out of rotten stumps or tree roots in deciduous forests. In taking the fruit body out of the substrate, the long root-like part of the stipe is usually torn off. *H. radicosum* is inedible. It must not be mistaken for *Hypholoma radicosum* which also has a rooting stipe but is more slender and strongly smells of iodoform.

Gymnopilus Sapineus

In summer and in autumn, clumps (rarely individual specimens) of conspicuous, mostly rusty brown gill-fungi fruit bodies may be found growing on putrescent stumps or dead trunks of conifers. These fungi usually belong to the genus *Gymnopilus* which was also separated from the no more valid genus *Flammula*. Fungi of the genus *Gymnopilus* have rusty, often bright yellow or orange brown colours turning black when exposed to alkalic solutions. Their stipe usually bears a ring.

Gymnopilus sapineus is substantially smaller than the rubust *G. spectabilis*. The cap is no more than 3-5 cm. wide, at first hemispherically convex, soon flatly expanded, with a dry cuticle, fibrous to finely brownish scaly on a golden-yellow background. The stipe is 4-7 cm. long, 5-10 mm. thick, cylindrical, often terminating in a rooting and whitish matted-tomentose base embedded in rotten wood. The stipe surface is silky shiny fibrous, pale yellow or only yellow-tinged. The gills are fairly crowded, adnate with tooth, at first pale yellow, then golden-yellow to vivid yellow, powdered with a rusty brown spore mass, slowly staining rusty where handled. The yellowish flesh has a bitter taste.

Several species closely related to *G. sapineus* have been described. Distinguishing between them is difficult, in fact it is possible only after detailed microscopic examination. None of these species is edible.

Spotted Tough Shank

The Spotted Tough Shank is distinguished from the other species of collybia by the size and fleshiness of its fruit bodies.

The 5-12 cm. wide cap is semiglobular, convex then expanded, with a narrowly involute margin; its surface is smooth, glabrous, originally white or slightly yellow-toned, sooner or later scattered with irregular brownish red, or rusty brown spots.

The stipe is often longer, than the cap diameter, 1-2 cm. thick, cylindrical, often flexible, fibrous tough and with an abrupt root.

The gills are very crowded, narrow, emarginate, attached by a tooth, white with finely serrate or minutely denate edges.

The spore print is yellowish-pink. The entire fruit body changes to reddish-brown with age or on handling. The flesh is white, tough, with a slightly sweetish smell and a tart, bitter taste.

The Spotted Tough Shank grows on rotten wood below ground, that is seemingly from the ground, sometimes directly on stumps and roots. It occurs exclusively in coniferous forests, especially in pine woods and in mountain spruce forests.

It appears mostly in late summer and in autumn, growing until the onset of forests. It is inedible.

Cystoderma Amianthinum

The species included in the genus *Cystoderma* are usually small fungi with a 2-6 cm. wide cap. In the past they were classified as members of the genus *Lepiota*. From these, however, they are distinguished by a granulose coating of both cap and stipe consisting of small conical warts, and by having gills adnate with tooth.

Cystoderma amianthinum has a slightly convex cap, ochre to deep golden-yellow, often wrinkled, densely granular. The gills are white when young, later ochre-yellowish, and crowded. The stipe is longer than the cap diameter, 2-5 mm. thick, cylindrical, with a slight clavate swelling at the base, ochre-yellow, bearing a scaly ring, with ochre-coloured scales below the ring. The flesh is yellowish, giving a faint rancid smell, its taste is mild. The spore print is pure white.

C. amianthinum grows from the end of summer until the first frosts in grassy habitats and in moss, particularly in damp coniferous forests, pine woods and heather moors, greatly favouring acid soils. It is inedible.

C. longisporum was distinguished from *C. amianthinum* as an independent species. Its cap and stipe are somewhat darker, reddish to rusty ochre, the gills are cream-yellow, sometimes pinkish. It is almost odourless. It is found in both coniferous and deciduous forests. Both species have amyloid spores. In the genus *Cystoderma,* examination of the amyloidity of spores is necessary before individual species can be exactly identified.

Volvariella Speciosa

The genus *Volvariella* consists of medium- to large-sized fungi with soft to watery flesh whose young fruit bodies are enveloped in a firm membrane; this breaks open as the stipe develops, leaving a clearly discernible, high sheath (volva) which often splits into lobes. The stipe is ringless, neither are there any visible remains of the veil on the cap surface in most cases. The broad, ventricose gills are free (as in the amanitas). The colour of the spore print substantiates the classification of *Volvariella* among red-spored gill fungi belonging—together with the genus *Pluteus*—to the family Pluteaceae. They somewhat resemble members of the genus *Amanita* in apperance.

Volvariella speciosa is one of the best-known representatives of its genus. It is a high and robust fungus having a 6-12 cm. wide, relatively fleshy cap with a thin margin, at first obtusely conico-campanulate, later expanded, slimy when moist, shiny, grey-whitish. If the colour of the cap is grey or blackish brown, especially in the centre, then this means that the variety *gloiocephala* is involved. The gills are crowded, whitish, and flesh-reddish when mature. The 20 cm. long and 1-1.5 cm. thick stipe is solid, white, slightly thickened to bulbous at the base, with a volva almost disappearing in older fruit bodies and remaining in the ground after the removal of the stipe. The flesh smells of rotten fruit.

V. speciosa is a typical fungus of gardens, parks, fields, urban rubbish heaps, occurring mostly on manured soils. It is inedible but not poisonous, as was previously thought.

Tephrocybe Atrata

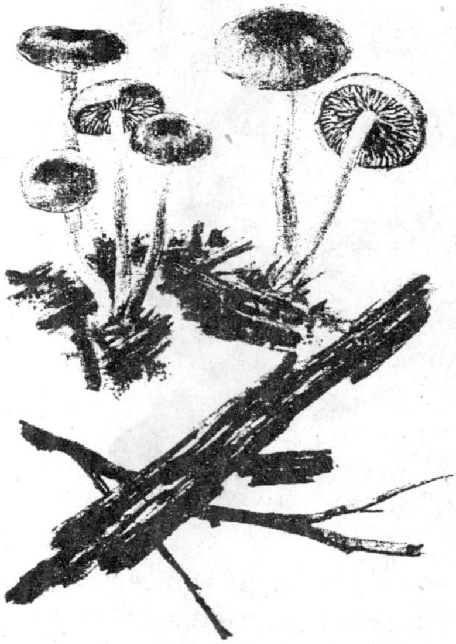

These are mostly smaller fungi with caps rarely exceeding 5 cm. in diameter, coloured in darker shades of grey, greyish-brown to blackish. Also the gills are more or less grey or off-white.

A number of species have an intense cucumber-mealy smell and a similar taste.

Tephrocybe atrata is one of the characteristic representatives of an ecologically interesting group of fungi whose occurrence is confined to burned ground only. It occurs, predominantly on burned-over areas an man-made sites and in forests.

Its cap is 2-4 cm. across, with relatively elastic and thin flesh, depressed at the centre, blackish-brown to black in wet weather, lighter in dry weather, and glabrous.

The stipe is thinly cylindrical, unequal, the same colour as the cap, whitish at the base, glabrous and cartilaginous.

The greyish gills are subdistant, broadly adnate, and subdecurrent. The crushed flesh smells strongly of cucumbers and flour with a tinge of cod-liver oil. The spore print is pure white. It is inedible.

Pholiota Carbonaria

On burned ground both within and outside the forest we may come across a great number of interesting

anthracophilous fungi whose occurrence is strictly confined to such an environment. One of the most common is *Pholiota carbonaria* which often covers the entire burned-over area with dozens of its fruit bodies.

The 1.5-6 cm. wide cap is at first semiglobular, then flat, with a narrowly incurved margin, thin-fleshed. The cuticle is sticky when moist, shiny when dry, smooth and glabrous, brown-yellow to cloudy reddish orange or rusty yellow. It is often contaminated with incrusting soil and charcoal particles. The stipe is usually equal to the cap diameter in length, relatively thin (2-5 mm.), pale yellow, whitish below the apex, finely scaly all along its length, covered in younger fruit bodies with fugacious white filaments of the cortina. The gills are moderately crowded, emarginate with tooth, pale or ochre-yellow when young, olive-brown in age, white ciliate at the edge. Numerous cylindrical, colourless cheilocystidia occur on the gill edges. The flesh is pale yellow, with a bitter taste and a faint acidulous smell. The spore print is rusty brown.

P. carbonaria is inedible. The fruit bodies grow singly or in sparse clusters, either upon burned wood or directly upon burned-over soil. *P. carbonaria* used to be classed into the genus *Flammula,* which no longer exists; the greater part of its species has been reclassified within the genus *Pholiota.*

Pholiota Destruens

Huge clumps of *Pholiota destruens* often appear on felled poplar trunks that remain lying on the ground for a long time.

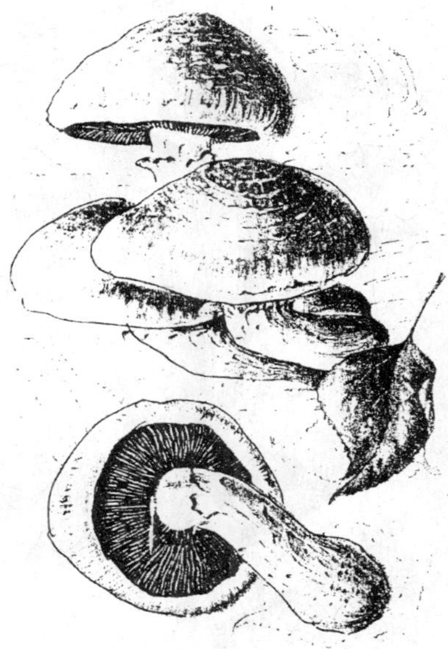

This is a very sturdy, firm-fleshed fungus; one of the largest in its genus. Its 5-20 cm. wide, dirty buff to pale brownish caps, convex in youth, later flat, are scattered with white scales. They are sticky when moist. The cap margin of young fruit bodies is joined to the stipe by a whitish veil, later fringed with its appendiculate fragments of irregular shape. The veil also leaves a fugacious ring on the stipe. The stipe is stout, pallid, scaly, 5-15 cm. long and 1-3 cm. thick, often inflated and rooting at the base. The brownish gills have a sinuate, irregularly denticulate edge.

P. destruens is inedible. Its bitter, rather tough flesh emits an unpleasant fungussy smell. Clumps of this *Pholiota* appear from August to November not only on dead trunks and stumps, but fairly often also directly on live trees, mainly on poplars and willows.

In the contemporary mycological system, the genus *Pholiota* is connected with the genus *Flammula,* which was once treated as independent. The only feature substantiating its separation from the genus *Pholiota* was its weakly developed, fugacious ring. Furthermore, the fruit bodies of most *Flammula* species are less fleshy and form smaller tufts than members of the genus *Pholiota*..

Brick Red Hypholoma, Bricktop, Bricktop Mushroom

The Brick Red Hypholoma usually grows in larger or smaller clumps, but solitary fruit bodies may occasionally be found—these, as a rule, are very robust. From the beginning of summer until the first frosts it can be seen growing out of

rotten stumps mainly of deciduous trees. Its taste is bitter, but is not so offensive as in the Sulphur Tuft. The flesh in the stipe apex is whitish, while in the Sulphur Tuft it is sulphur-yellow in the whole stipe. The cap of the Brick Red Hypholoma is yellowish red in the centre, sometimes almost brick-red, lighter coloured at the margin where it changes to orange or yellow. It is 3-10 cm. wide and when young, is covered with scaly or fibrous cobwebby remnants of a whitish or yellowish veil connecting the cap margin with the stipe in very young fruit bodies. The remnants of the veil are subsequently washed away by rain or disappear in the course of the fruit bodies' development; thus the cap surface of mature fungi is glabrous. The 7-12 mm. thick stipe is usually firm-fleshed, later hollow, yellowish, rusty brown at the base, fibrous scaly, with veil remnants in the form of whitish or yellowish filaments finely powdered with loosened blackish grey spores. The gills are crowded, adnate with a tooth, at first yellowish, then yellowish brown, mature gills are olive- to blackish-brown. The spore print is olive dark brown.

The Brick Red Hypholoma is not recommended for table use, not only because of its bitter taste but also because of possible confusion with the Sulphur Tuft.

Tricholoma Vaccinum

Tricholoma vaccinum is a small to medium-sized fungus with a typical reddish colouring, scaly cap surface and bitter taste.

The cap is 3-9 cm. wide, rarely larger, relatively thin fleshed, and of red brown or reddish rusty colour covered with fibrous erect scales, all over the surface.

The cuticle between the scales is lighter and fibrous felted. The stipe is up to 10 cm. long and upto 1.5 cm. thick, cylindrical sometimes slightly inflated at the base always hollow. The stipe surface is fibrous scaly, and the same colour as the cap.

Below the apex of the stipe under the gills there is a sharp colour divide the stipe apex being lighter coloured than the rest.

The fibrillose veil which in the young fungus connects the cap margin with the stipe soon disappears.

The gills are emarginate and decurrent by a tooth, dirty whitish later dingy reddish, and up to 15 mm. broad. The flesh is whitish then reddish or with a rusty tinge; it has an earthy smell and a bitter-sour taste. The spore print is white.

It is unsuitable for eating because of its bitter taste.

Geastrum Triplex

There are approximately 30 species of earth stars (*Geastrum*) growing in Europe, and they are among the true gems of nature. They occur very rarely however, and are rapidly disappearing due to mankind's devastation of the countryside.

Geastrum triplex is one of the largest earth-star species. When fresh, it is also one of the most beautiful. It usually forms sparse colonies in humus-rich woods, in gorges, around old stumps, and in shady localities. Though it prefers deciduous forests, it also occurs in unraked and unkept remote parts of old gardens and parks, especially under oaks, ashes, beeches, hornbeams, maples and limes. Sometimes it also makes its appearance in raspberry and nettle growths, and on rubbish heaps in woods. It is rare in coniferous forests, occurring especially under old spruces or firs in various herb-growths. It grows from September till November. Young, still-closed fruit bodies are onion-shaped, ochre-brown to pale pinkish. The exoperidium bursts into 4-7 (mostly 5) pointed, unequal rays; the endoperidium is sessile, with a fibrillose mouth delimited by a clearly discernible circular area. The specific name *triplex* is derived from the apparent division of some fruit bodies into three parts (the endoperidium, the collar arisen from the flaring fleshy part of the cracked exoperidium, and the incurved rays of the exoperidium at the base). *G. triplex* is inedible.

Astraeus Hygrometricus

Fungi with star-shaped fruit bodies have always attracted the attention not only of laymen but also of mycologists. This is also why *Astraeus hygrometricus* has usually been subjected to study in connection with the order

Geastrales, though the genus *Astraeus* bears no affinity with Earth Stars (*Geastrum*). *Astraeus* belongs to the independent family of Astraeaceae included in the order Sclerodermatales; this family is more closely related to the genera *Scleroderma, Pisolithus* and *Tulostoma.*

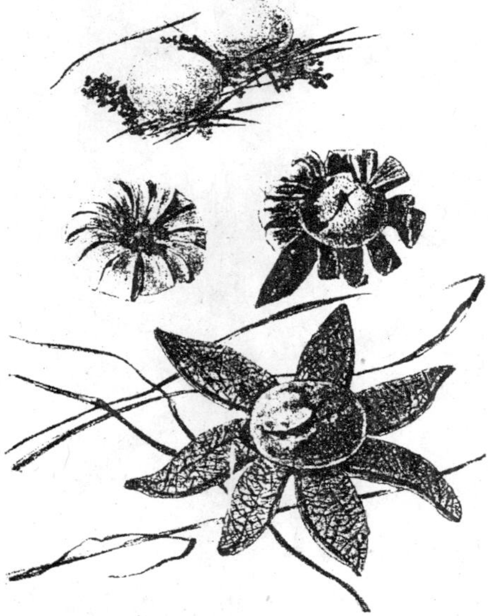

Seated in the centre of the coriaceous outer peridium, which splits at maturity to form regular stellate rays, is the globular inner peridium with an irregular, small peristome at the apex. The rays of the outer peridium tend to curve inward or coil up in dry conditions (so covering the inner peridium), and to expand again in rainy weather. In old fruit bodies the surface of the rays is densely cracked into minute areoles; they are dark grey to black in colour. The fruit bodies remain in their habitat for a very long time—often for many years.

A. hygrometricus grows in colonies on sunny rocks, on stony and rocky slopes overgrown with sparse woods, and in the outskirts of oak or pine forests. It occurs predominantly on more acid substrates (on slate, granite and sands) and avoids lime-rich soils. Fresh fruit bodies are found from autumn till spring, older ones throughout the year. It accompanies almost exclusively oaks and pines, and its distribution is world-wide. It is inedible.

Sphaerobolus Stellatus

This is one of the smallest of the Gasteromycetes. When looking for it, one must carefully examine old and decaying plant remains such as stems, branches, pieces of wood and

bark, but also old cow dung and excrements from other herbivores. On worked wood exposed to enduring humidity. e.g. in hothouses, hotbeds, etc., *Sphaerobolus stellatus* can even manifest itself as a wood-destroying fungus—otherwise it is of no particular importance.

The development of its fruit bodies is an interesting sight; from the tiny globular fruit bodies, only 1-4 mm. in size and white-felted on the outside, an equally globular bright orange pip (peridiole) is released and discharged to a considerable distance under strong pressure. The emptied fruit body has the shape of a deep, semiglobular saucer with a margin split to form 5-8 pointed rays, neatly star-like. Reproduction is effected by the discharged globular peridiole containing basidia with spores.

S. stellatus is most commonly found after summer rains. We can bring its fruit bodies home together with the substrate, and watch the jerking out of the pips stirred up by sudden illumination or another impulse (such as touch).

11

Poisonous and Deadly Poisonous Fungi

Nowadays a responsible evaluation of the edibility or toxicity of many fungus species is often problematic, since it comes to light that species formerly currently collected eaten are less innocent than was previously assumed.

Involved here are not only the toxic principles representing the direct cause of poisoning but also other, largely still unknown toxins which can accumulate in the human body after repeated ingestion of some mushrooms, and unfavourably interfere, with the activity of various organs, mainly of those belonging to the digestive system.

Numerous fungus species contain substances detrimental to human beings. In the past most fungi were considered poisonous or at least viewed with suspicion and people preferred to shun them without thinking of using them for food. Later on, however hand in hand with a more thorough acquaintance with species and progressing mycological knowledge, the assortment of edible and consumable species started to expand.

Mushrooming was ardently indulged in several European countries. The popularization of mushrooming was sometimes so inordinate that almost all fungi were declared edible.

It was only the growing incidence of cases of poisoning of cases of poisoning, which frequently proved fatal that provided a warning signal and again called for increased caution.

According to the symptoms or types of toxins concerned poisoning can roughly be divided into several categories. In each of these we shall confine ourselves to those species of poisonous fungi which come into account, as the most frequent cause of poisoning.

Phalloid Poisoning

Poisoning that damages the liver is called phalloid poisoning. This name is derived from the specific name of the Death Cap.

Amanita phalloides—which is the most frequent cause of this type of poisoning.

The fruit body of the Death Cap contains two groups of poisonous substances, present in high concentration, namely amatoxins and phallotoxins. Both groups of toxins are also present in the closely related species of the, so called poisonous, white amanitas, in *A. verna* and *A. virosa,* in concentrations still higher than in the Death Cap.

Anybody intending to collect edible mushrooms and not willing to confine himself to gathering boleti and chanterelles must inevitably be acquainted with the Death Cap. Although the Death Cap is greatly variable in colour, some of its characteristics are so conspicuous that it can readily be recognised even in case of an atypical colouring. Its toxicity manifests itself after the ingestion of quite a small amount of food prepared of it. One fully developed fruit body was found capable of causing the death of two persons. The first symptoms appear after a relatively long interval (the so-called period of latency), six hours after the ingestion at the earliest (earlier after consuming a raw toadstool), usually in 8-24 hours; sometimes the period of latency is still longer. The first symptoms of poisoning appear as an irrigation of the digestive systems—intensive vomiting and diarrhoea. These symptoms may last for 48 hours in the course of which the patient suffers great losses of body fluids and mineral substances followed by extreme exhaustion. Cramps in the legs subsequently set in, accompanied by urination disorders which may result in a total stoppage of urine secretion and generally apathy, since the organism suffers from poisoning by waste materials it fails to get rid of. The disappearance of diarrhoea and vomiting signals the onset of a short but extremely dangerous period of alleviation which, however, only precedes a further phase characterized by liver-tissue impairment. This period is critical. If sufficiently effective help has not been secured in time, or if the afflicted person has eaten a greater number of fruit bodies, he becomes unconscious and dies on the fifth or sixth day with symptoms of liver failure and sometimes of brain tumescence.

Thirty years ago, up to 80 per cent of those poisoned died after having eaten the Death Cap. Nowadays, when new therapeutic methods are continuously being applied in departments of intensive care, the death rate has considerably fallen and does not exceed 15 per cent. At any rate, phalloid poisoning always seriously endangers life. The non-existence of such an acid, pointing out that a mixture of several non-toxic substances is involved. Simultaneously, however, a new effective principle was isolated and called gyromitrin, which quickly decomposes into the strongly poisonous methylhydrazine. However, both the mentioned substances are extremely volatile and disappear almost completely after 10 minutes' cooking. After a normal heat-preparation of fresh *G. esculenta,* or table use, neither gyromitrin, nor methylhydrazine can be regarded as the cause of poisoning. In this fungus further four substances related to gyromitrin have been discovered; of these, hemolysin initiating the lysis of red blood cells is also destroyed by cooking. The most plausible explanation is that some old and infected fruit bodies of *G. esculenta* become poisonous only secondarily.

Occasional cases of poisoning after eating some other larger Discomycetes—*Gyromitra gigas, G. infula, Sarcosphaera crassa,* etc.—take the same course as poisoning by *G. esculenta.* Both gyromitrin and methylhydrazine were found in some of them.

Cases of poisoning by the Sulphur Tuft have repeatedly been reported from Russia as well as from some central and southern European countries. Their course was typically phalloid. Japanese mycological literature states that the Sulphur Tuft belongs to poisonous fungi responsible for the greatest amount of poisoning in Japan. This is almost incomprehensible in view of the fact that the Sulphur Tuft has an offensively bitter taste which remains unaffected by cooking.

Muscarine poisoning. Its name is derived from the specific name of the Fly Agaric—*Amanita muscaria*. In the past, fruit bodies of the Fly Agaric were steeped in milk used for poisoning flies. It has been discovered only in the present century that both the Fly Agaric and the Panther Cap (*A. pantherina*) contain but a small amount of muscarine, and that their toxicity is due to another toxic principle. On the other hand, muscarine was found in large quantities in other species, particularly in those of the genus *Inocybe* and in some white species of *Clitocybe*.

Muscarine poisoning sometimes manifests itself while the meal is still being eaten—as a rule, however, within two hours after it. The symptoms are conspicuous perspiration, salivation, lacrimation, vomiting and diarrhoea, fall of blood pressure, slowing down of the pulse and shivering. A remarkable symptom is the contraction of pupils accompanied by sight disorders. A reliable antidote is atropine; this, however, must not be applied in poisoning caused by the Fly Agaric and Panther Cap. In the whole of Europe, muscarine poisoning is mostly caused by *Inocybe patouillardii* containing approximately 500 times more muscarine than the Fly Agaric. A fatal dose is contained in 100-150 g of a fresh fungus. Also other *Inocybe* species contain relatively large quantities of muscarine—e.g. *I. lacera, I. geophylla, I. argillacea,* and it is only in some species of this genus that no presence of muscarine has been proved (e.g. in *I. jurana*). Also several white *Clitocybe* species contain large quantities of muscarine—namely *C. dealbata* and the closely related (or perhaps identical) *C. rivulosa, C. phyllophila,* and *C. cerussata.* A small amount of muscarine was discovered in some *Hebeloma* species, e.g. in *H. crustuliniforme* and *H. sinapizans,* but also in *Armillaria mellea* and *Omphalotus olearius.* Slight symptoms of muscarine poisoning appear after their ingestion. Moreover, *O. olearius,* abundant in southern Europe, causes symptoms showing the presence of another content principle, hitherto unknown. According to some authors the true *O. olearius* occurs only in southernmost. Europe, while the fungus growing further northwards (e.g. in the Czech Republic) belongs to a different species, probably related to some poisonous *Omphalotus* species from North America, containing the toxins illudin S. and M. Also the type of poisoning caused by *Mycena rosea* bears some resemblance to muscarine poisoning; its content principle has not yet been discovered.

Psychotropic poisoning involves serious cases characterised by the irritation of brain tissue. For a long time the intoxication caused by the Fly Agaric was the only form of mushroom poisoning accompanied by psychic disturbances. It was not before the 1950s that other so-called cult fungi, formerly used in religious ceremonies and rites, were identified; their ingestion leads to different manifestations of psychic disturbance. Two types of psychotropic poisoning are distinguished: psychotonic poisoning caused by the so-called mycoatropine, and psychodysleptic poisoning caused by psilocybine.

In Europe, poisoning by mycoatropine is caused by three *Amanita* species. Most common are cases of poisoning after eating the Panther Cap, less frequent are those caused by the Fly Agaric, and practically unknown is poisoning by *A. regalis.* The poisonous content principles of these amanitas have not yet been exactly identified, and this is why the designation 'mycoatropine poisoning', though inadequate, is still used nowadays.

The course of poisoning caused by all the three species is substantially the same: nausea is experienced between half an hour and three hours after consumption, accompanied by vomiting, headache, quickened heartbeat, and a persistent dilatation of pupils occasionally leading to vision disturbances. Often the condition of the affected persons resembles alcoholic intoxication: the patient becomes talkative, shouts obscenities, sometimes laughs or weeps, strikes himself and keeps on running to and fro. These states of excitement may be dangerous for the sick person and must therefore be mitigated. Subsequently the and mineral substances), possibly resulting in kidney damage. The same symptoms may be observed in poisoning by *Entoloma lividum,* eventually by other poisonous *Entoloma* species—such as *E. rhodopodium* and *E. nidorosum.* Poisoning caused by *Nolanea verna* is accompanied by an intensive diarrhoea lasting for several days and by the consequent exhaustion of the organism. Here the first symptoms of poisoning appear in 2-4 hours, maybe also in 24 hours. In some countries there have been cases of poisoning by *Agaricus xanthoderma,*. because of its confusion with edible agarics of a similar appearance. It is only after eating a large amount of fruit bodies that this type of poisoning manifests itself by persistent vomiting which may last for several hours. Also after consuming large amounts of fruit bodies of fungi sometimes used as spices, digestion disorders may turn up—for example in case of *Lactarius helvus, Collybia fusipes, Scleroderma citrinum* and *S. verrucosum.* In the case of S. *verrucosum* symptoms of poisoning appear very quickly after eating this mushroom fried on fat.

In specialized literature, data concerning the edibility or inedibility of many fungus species are often contradictory. In general, increasing knowledge tends to increase the number of species considered inedible or poisonous, rather than the other way round. Hence it is urgently recommended that the collection of mushrooms for table use should be confined to the narrower assortment of well-tried species to avoid running any risk. The rule of distinguishing edible species of *Russula* and *Lactarius* according to the taste of the flesh of fresh fruit bodies helps to exclude all the hot-tasting fungi belonging to these genera as inedible, although some of them are suitable for eating if subjected to a special preparation (this actually happens in some countries). Hot-tasting *Russula* and *Lactarius* species (e.g. *Lactarius turpis* and *L. torminosus*) may cause poisoning which manifests itself by stomach and intestinal troubles. This is caused by substances of a still unknown chemical composition characterized by thermolability. This is why in some countries, particularly in the north, these mushrooms are collected, cooked and preserved in salt or fermented. However, in southern and central Europe the very same species are considered inedible or poisonous.

In conclusion let us briefly refer to mushrooms currently collected as edible whose ingestion, however, sometimes calls forth unpleasant troubles or individual intolerance. Cases of this kind have so far not been explained, and naturally the content principles causing them are still unknown. For example, poisoning by the Honey Fungus whose symptoms appear after eating not only raw or inadequately prepared mushrooms but also well-prepared ones is completely unaccountable. It is assumed that a poisonous species may exist as one of the complex of so-called microspecies (the Honey Fungus represents exactly such a complex) which have not yet been satisfactorily distinguished. Also the brown-coloured *Tricholoma* species, especially those growing in coniferous forests (mostly under pine trees) on the lower slopes of mountains, include suspect species calling forth symptoms of poisoning. Some yellow or greenish species of *Tricholoma,* often bitterish to burning in taste, are mistaken for *T. flavovirens* and cause gastric troubles. The separate species of the genus *Ramaria,* abundantly collected in many regions, are not distinguished by practical mushrooms gatherers; they contain the laxative emodin, concentrated in the branch tips of their fruit bodies. It is therefore necessary to use them with care. Allergenic properties may be observed in many dozens of other fungus species described in books as being edible, and are mostly eaten without ill-effects.

It is a generally known fact that no alcoholic drinks must be taken after eating meals prepared with certain mushrooms species, otherwise symptoms of poisoning might appear. Best known in this respect is *Coprinus atramentarius.* In most cases, poisoning makes itself felt soon after ingestion; its symptoms include reddening of the skin, intense heart palpitation, asthmatic troubles, vomiting, diarrhoea and feelings of anxiety. They disappear within two hours but can re-emerge after drinking some more alcohol. The same holds for some other mushrooms as well.

12

The Morphology of Poisonous Fungi

As we have studied in the earlier chapter, numerous fungus species contain substances detrimental to human beings.

Now let us study about the morphology of some poisonous Fungi. They differ from each other, in their external structure as well as the, substance content.

Some fungi may exhibit narrow body part, where as the other, wide cap, and fibrous gills. They also differ in their habitat; now let us go through certain poisonous and deadly poisonous Fungi.

Amanita Umbrinolutea

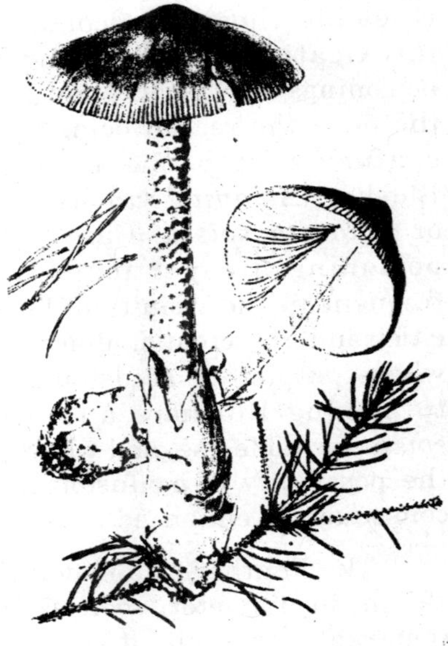

Amanita umbrinolutea is closly related and similar to the Grisette whose variety, it is sometimes considered to be. It is distinguished from the latter by its colour and larger spores.

The 5-12 cm. wide cap is at first avoid and enclosed in a firm membranous veil, later it becomes slightly convex to flatly expanded, with an attenuated and distinctly grooved margin.

The universal veil adheres to the apex of the young fruit body's cap but slough's off later, leaving the cap surface of older specimens glabrous.

The cuticle is slimy at first, then dry shiny, yellow-brown and whitish at the grooved margin.

The gills are free, loosely packed, thin and white. The cylindrical stipe is white, typically streaked with grey to yellowish grey. The ring is absent as is the case in all, vaginaria. The spore print is white.

It grows in the grass of woodland meadows. Its colouring sometimes resembles that of *A. fulva,* whose cap is reddish brown, paler towards the margin.

All the amanita species are poisonous.

Panther Cap

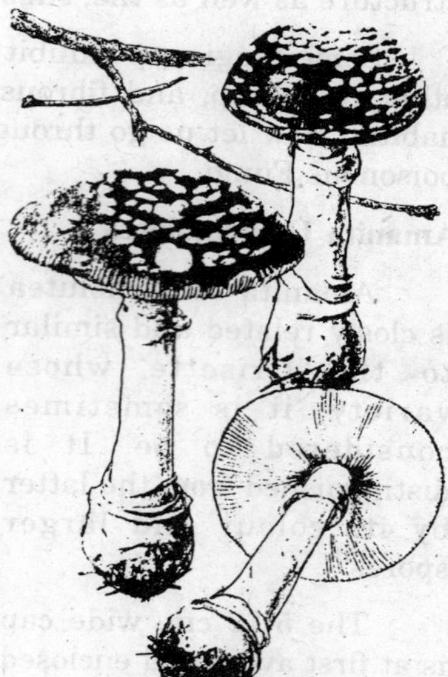

After the Death Cap, the second most frequent cause of poisoning is the Panther Cap. The symptoms of poisoning resemble those of alcoholic intoxication and more poisonings now occur than in the past, the reason being its confusion with the edible Blusher (*Amanita rubescens*) or *Amanita spissa.* Cases of poisoning are particularly frequent in the vicinity of big cities and in recreational areas where people are accustomed to picking Blushers and *A. spissa* for table use, and where the possibility of confusion is consequently enhanced.

The Panther Cap is not difficult to identify. Its cap is 4-12 cm. in diameter, soon expanded. Its colour variability is remarkable—usually it is yellowish brown but it may also be beige, greyish or grey-brown. Its surface is slimy when moist, shiny when dry, with a distinctly grooved margin, usually strewn within small fragments of the universal veil. These fragments are white and often concentrically arranged, thus vaguely resembling a panther's skin: hence the name of the toadstool. The gills are crowded and cream-yellowish. The stipe, up to 15 cm. long and 2 cm. thick, is white, soon hollow, with a white, smooth, narrow, ungrooved, often fugacious superior ring; it is slightly tuberously enlarged at the base. The tuberous base has a shallow ridge on the upper angle; above this, 1-3 slanting rings are often situated. The white, unchanging flesh has an odour and taste of raw potatoes and smells sweet when drying. The Panther Cap grows in summer and in autumn mostly in deciduous forests, usually under oaks. The ungrooved ring and the context exhibiting no colour change are characteristic attributes of the Panther Cap.

Fly Agaric

The Fly Agaric (1) has been known as a poisonous species since ancient times. Its toxicity is mainly due to the presence of mycoatropine which causes disorders of mental activity. The content of another poisonous principle, muscarine, is relatively small. Recently the identity of the Fly Agaric with the drug called 'soma', venerated by the most ancient Aryan tribes in the time of migrating to and settling in the mountains of Afghanistan, has been established. The migration of peoples contributed to the further spreading of the Fly-Agaric cult. Particularly

remarkable is the Siberian cult of the Fly Agaric: people were drinking fruit-body decoctions, chewing dry toadstools and washing them down with cold water, or they would prepare a beverage from a mixture of the toadstool and leaves of the Bog Whortle-berry nad *Sadix angustifolia*. Since the effective substance is secreted with urine, they even drank the urine of intoxicated person.

The symptoms of poisoning include vomiting, headache, accelerated heartbeat, dilatation of pupils; often a state similar to alcoholic intoxication and hallucinations set in, and finally the poisoned person falls into a deep sleep and usually awakes in the morning in a normal condition, without remembering his or her previous behaviour.

Amanita regalis (2), growing in upland spruce stands, is distinguished by a yellowish brown cap, a yellowish stipe and similarly coloured remnants of the outer veil on the cap, and by a ring. It seems to be as poisonous as the Fly Agaric.

Grisette

Not all species of *Amanita* have a ringed stipe—in many of them the membranous universal veil (*velum universale*) persists in the form of a volva, i.e. a sheath which envelops the base of the stipe and roughly resembles half of an egg-shell. In the past, these ringless *Amanita* species possessing a volva were classified as members of the genus *Amanitopsis*, yet the present mycological system treats them as a mere section (*Vaginaria*) of the genus *Amanita*.

The Grisette and several other closely related species are the most common ones. The typical form has a brittle, thin-fleshed cap, light or dark grey, frequently fading to off-white. It is usually 3-12 cm. wide, soon flatly expanded, slightly and obtusely umbonate, with a markedly grooved margin. The stipe and the volva are whitish, the gills are pure white, crowded, and free. The spore print is white.

The Grisette occurs throughout the summer until the autumn in all types of forests. It is mostly found on damp soils overgrown with grass and moss, often under birches growing on acid substrates. It is poisonous when raw (as are all the species of this family) and is easily confused with the extremely poisonous Death Cap.

The related *A. fulva* closely resembles the Grisette in appearance and size, except for its reddish brown cap and an orange-yellow volva. A common and quite abundant species, it favours spruce stands rich in moss but also grows under deciduous trees.

Inocybe Fastigiata

This is one of the most abundant species of Inocybe. It is extremely variable in colour. One of its distinguishing features is an acutely conical cap, 2-8 cm. wide, with a conical umbo, persisting even later, after the cap has become flatly expanded.

The cuticle is markedly radially cracked, ochre-yellow, yellow-brown to dark brown. The gills are free, ventricose, light olive-yellowish gradually turning brown, whitish in the edges.

The cylindrical often considerably elongate, 3-8 cm. thick stipe is whitish or pale yellow. The flesh is white unchanging with supermatic smell and an unpleasant taste.

The spore print is ochre-brown. Colourless vesicular cheilocystidia develop on the gill edges.

Inocybe fastigiata grows in both deciduous and coniferous forests, especially on bare ground by tracks or on ditch banks, from summer until autumn.

It is poisonous, the fruit bodies contain a large amount of muscarine. Its variety designated as *Var. umbrinella,* has a dark greyish brown to umber-brown coloured cap.

Inocybe Patouillardii

With regard to the number of species included, the genus *Inocybe* is one of the richest genera of gill fungi. Well over a hundred species grow in Europe. They are of considerable practical importance; many of the species are poisonous fungi—some of them are highly toxic. This also applies to *Inocybe patouillardii* which is all the more dangerous as its large, fleshy fruit bodies make a pleasant sight and attract mushroom pickers at a time when the edible St. George's Mushroom (*Calocybe gambosa*) appears in similar localities. Consequently cases of poisoning are reported every year.

The fruit bodies of *I. patouillardii* have caps up to 10 cm. wide, conically bell-shaped at first, later flatly expanded, umbonate in the centre, with a usually wavy-lobed margin. The silky fibrous cap cuticle is at first white or whitish, but fairly soon—especially when handled—develops brick-red stains. This rubescence is very characteristic of *I. patouillardii:* it can be seen not only on the cap but also on the stipe and gills. Also the originally white flesh, when cut, turns red in the stipe and pink in the cap. Older fruit bodies are entirely brown-red to brown. The stipe is cylindrical, up to 12 cm. long, 1-1.5 cm. thick, swollen at the base. The gills are at first white, then olive-brownish. The spore print is brown. Cylindrical to claviform, glabrous cheilocystidia are situated at the gill edges. The flesh has a faint fruity smell and a mild taste.

I. patouillardii grows from May onward, but especially in June and July under deciduous trees, often in parks and hedgerows, on lime-rich soils. It is deadly poisonous, containing five hundred times more muscarine than the Fly Agaric.

Verdigris Agaric

Some *Stropharia* species resemble small field agarics, since they have in common the colour of the spore print (brown or violet-black) and a veil leaving a ring on the stipe.

The most commonly occurring species of *Stropharia* is *S. coronilla*. Its cap is up to 6 cm. in diameter, yellowish white to pale ochre. The gills are chocolate-brown, the stipe is relatively short and bears a grooved ring. It occurs after rainfall in dry lawns, pastures and along field margins.

One of the strikingly coloured species is the

Verdigris Agaric whose 3-10 cm. wide, green cap, weathering to ochraceous later, is covered with a layer of slime. The green or green-blue stipe is also slimy, 4-8 cm. long and 0.3-0.6 cm. thick, white scaly, bearing a collar-shaped ring below the apex. The blue-greenish flesh emits a peculiar tannin-like smell, or one rather like rotting radish, which—together with the slimy cuticle—does not make the Verdigris Agaric particularly attractive for mushroom pickers. Until recently it had been considered edible; now, however, it is classed as poisonous and should be avoided.

Fruit bodies of the Verdigris Agaric grow from August to November on the remains of wood logs and branches lying on the ground or hidden immediately beneath the surface. Sometimes they also grow on rotten stumps of deciduous and coniferous trees, especially in mixed forests, gardens, fields, parks, shrub growth, etc.; they may even occur on rotting straw.

Blue-green is also the characteristic colour of the *Clitocybe odora;* this, however, has no veil, smells of aniseed, and its spore print is white.

Inocybe Lacera

This is the earliest species of Inocybe appearing as early as May on bare forest ground, along roadsides, in woodland glades and in clearings, in rather dry sun exposed places, mostly in coniferous forests on acid substrates. It is most abundantly found in heather moors or in sandy pine woods, where it can be quite common until autumn.

The uniform rather dark brown colour of its cap and stipe serves as a satisfactory diagnostic feature.

The cap is 2-5 cm. wide with a soft fibrous-felted surface scaly around the apex. The stipe is rather long and slender, 2-4 mm. thick, relatively firm, fibrous all over its earthy grey-brown surface. The gills are crowded, almost free, ventricose, whitish at first then pale earthy, finally brown with white ciliate edges.

The flesh is whitish, thin in the cap brownish in the stipe with a rather unpleasant odour, the spore print is pale brown.

Though greatly variable it can always be safely recognized by characteristic spores.

Hygrocybe Nigrescens

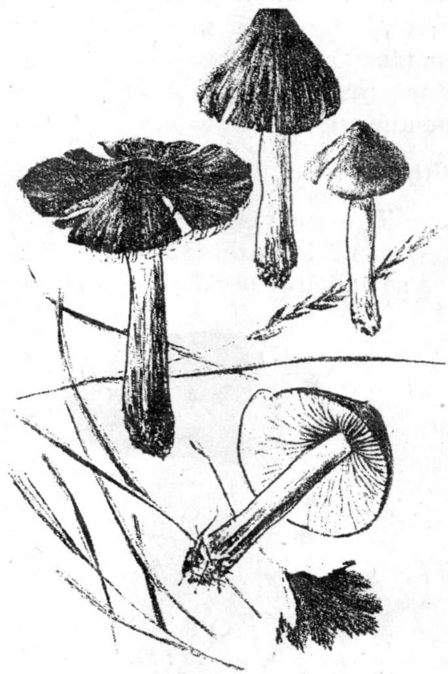

In the colouring of the genus *Hygrocybe* the predominant colour shades are bright yellow, orange and red; in some species these are mutually combined and complemented by still other colours. Fungi belonging to this genus ar readily distinguished by their colouring and a peculiar, almost waxy, consistency of their fruit bodies. These wonderful little fungi are usually found outside the forest, in grass and moss—mostly in natural localities substantially unaffected by man's interference. However, because such localities are disappearing very rapidly, the species of *Hygrocybe* are increasingly becoming a rarity. The fruit bodies appear in summer or in autumn after heavy rainfalls. Their vitality is low owing to their exposure to sunshine, which is usually intense outside the forest.

Hygrocybe nigrescens and its double, *H. conica,* are among the most common and, due to their peculiar colour change, most conspicuous of *Hygrocybe* species. These two species have one characteristic in common; they gradually blacken on injury or with age. Young specimens have vivid colours and a remarkable shape. The cap is 2-6 cm. across, acutely conical, tapering to a point, orange-yellow or scarlet-red, with a radially fibrous cuticle; the gills are yellow, the stipe is yellow or orange, longitudinally fibrillose-striate. The differences existing between these two species are negligible. *H. conica* is smaller, its cap is only 2-4 cm. wide and sharply conical, its stipe is orange rather than lemon-yellow, its basidia are bisporous. *H. nigrescens* is more robust, with a more pronouncedly lemon-yellow stipe and tetrasporous basidia. Both species are slightly poisonous.

Bitter Boletus

The greyish pink spore print is one of the main diagnostic features differentiating the genus *Tylopilus* from the so-called true boleti of the genus *Boletus*. The 'true' boleti have a brown or olive-brown spore print. Most of the species included in the genus *Tylopilus* (about 10 of which are known) do not form any mycorrhizal associations but grow as saprophytes on forest humus and very putrescent wood. In Europe this genus is represented by one species only, i.e. by the Bitter Boletus—most of the others are widespread in North America.

As a rule, the Bitter Boletus grows in coniferous (mainly spruce and pine) forests on acid soils, both in lowlands and in the mountains. It often grows on decaying stumps overgrown with moss. In the past decade it has been expanding its range, frequently flooding the forests as a weed fungus. Its neat fruit bodies strongly resemble those of the Edible Boletus but their offensive taste makes them uneatable. Careless or inexperienced mushrooms pickers are disappointed by the food prepared with these fungi: their distasteful bitterness cannot be eliminated by any processing. Nonetheless, the Bitter Boletus may readily be distinguished from true boleti when picking them—not only by their bitter taste but, even without tasting them, primarily by their greyish pink or flsh-pink tubes, mostly conspicuously pulvinate (this applies especially to older fruit bodies). Also the reticulum covering the stipe is substantially coarser, with meshes larger and deeper than in the Edible Boletus.

Livid Entoloma

Fruit bodies of this attractive fungus develop at the beginning of July in oak forests on calcareous soils and continue to grow until September.

They may occur also outside the forest under solitary oaks on pond dams or in tree alleys. A thermophilous species, it avoids cool and mountainous areas and never grows on acid substrates.

Sometimes it occurs together with Boletus aestivalis and the young fruit bodies, resembling young boleti in shape, can be gathered together with them by inexperienced mushroom pickers.

The fruit bodies growing in early summer can be mistaken for belated St. George's Mushrooms.

The ingestion of the extremely poisonous Livid Entoloma causes vomiting and diarrhoea in about 2-4 hours; this conditions may last for several days. The toxic principle of a hitherto unknown composition can be destroyed neither by boiling nor by the use of vinegar in preservation.

The caps of the robust fleshy fruit bodies are 5-15 cm. wide, at first semiglobular with their margin pressed against the stipe.

The flesh is white with a cucumber-mealy smell and taste. The spore print is reddish.

Nolanea Verna

When the weather is favourable, the first fruit bodies of *Nolanea verna* appear in early spring in sunlit, grassy woodland sites. Characteristic is their dark colour and greyish pink gills. The 3-6 cm. wide cap is conical or bell-shaped, later slightly convex, umbonate in the centre, fragile, brittle, strongly hygrophanous, deep dark brown to black-brown when moist, pale grey-brown with a silky sheen when dry. The gills are distant, broad, grey when young, grey-pink and covered with pink spore powder at maturity. The stipe, sometimes exceeding the diameter of the cap in length and 0.5 cm. thick, is laterally depressed, brittle, hollow, and grey. The flesh is greyish, almost odourless and tasteless.

N. verna is an extremely poisonous fungus whose ingestion is followed by poisoning with diarrhoea as its main symptom. Careless or inexperienced mushroom pickers usually take *N. verna* instead of the Fairly Ring Fungus because it also grows in grass. Yet it does not resemble in any way the Fairy Ring Fungus, being absolutely dissimilar not only in it colouring and fragility, but particularly in the pink or fleshy red spore mass and angular spores. There is a multitude of gill fungi with spores of this colour, and many of them are demonstrably poisonous.

Another vernal species is the Roman Shield Entoloma (*Entoloma clypeatum*). It also grows in grass under some fruit trees, mostly under plums. It is much larger and more fleshy than *N. verna* and has a white stipe; its flesh emits a farinaceous smell. It is edible.

Lilac Mycena

Some *Mycena* species can be recognized by their smell. This applies also to be Lilac Mycena with its strong smell of radish. This species varies greatly in size and particularly in colour. In its typical form the cap is pale to rosy violet but there are also bluish, brown and white varieties. In a drying fruit body the intensity of colouring will change: it is always paler than when moist. The tubular stipe is brittle, covered with long white hairs in the basal part, 3-7 cm. long and 0.3-0.5 cm. thick. The distant, broad, deeply emarginate gills are interveined at the base, and are pale violet.

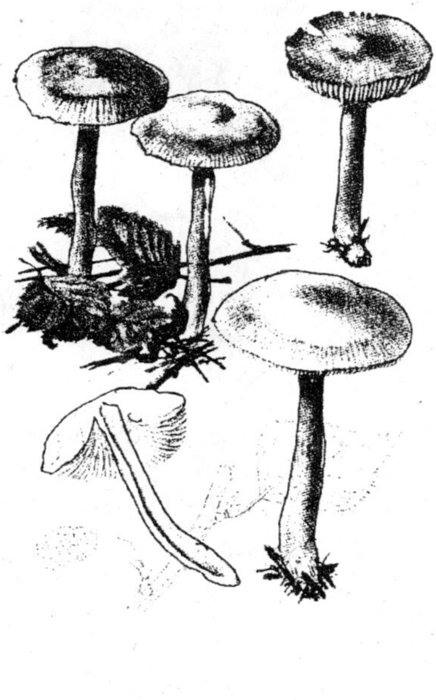

The Lilac Mycena grows in most various types of forests from May until October. Both solitary and groups of fruit bodies develop in humus and detritus, rotting foliage and needless, always favouring damp situations. Older data concerning the toxicity of the Lilac Mycena have recently been revised: today it is considered definitely as poisonous. Since the symptoms of poisoning following the ingestion of a large amount of fruit bodies are not unlike those caused by the Fly Agaric, the Lilac Mycena presumably contains the poisonous alkaloid muscarine.

It is practically impossible to confuse it with any edible violet-coloured gill fungi. For instance, the Wood Blewitt is a fleshy, robust mushroom of quite a different appearance, with a pale pink spore print. *Laccaria amethystina* is recognizable by its tough stipe and thickened gills strewn with white powdery spores.

Lorchel, Turban Fungus

The Lorchel is a typical representative of the vernal mycoflora of coniferous, particularly pine forests. Here its fruit bodies appear in May and at the beginning of June, usually on bare ground.

It prefers open spaces e.g. woodland clearings where it can be found near to stumps; sometimes it also thrives directly on these or on their roots.

The Lorchel is not recommended for eating, or is forthrightly classed among the poisonous fungi.

Poisoning by the Lorchel is somewhat puzzling. Although young fruit bodies are edible without any ill effects, cases of poisoning have been reported, even if rarely—after the ingestion of older fruit bodies.

This is why in some countries its sale has been for bidden. The substance causing this poisoning as well as its mechanism have so far eluded explanation.

Lactarius Helvus

In moist coniferous forests overgrown with moss, especially on cid substrates, whole swarms of this ochre-brown *Lactarius* commonly appear in summer and in autumn. It has a particular liking for soils saturated with water and is often found in peat-bogs among live *Sphagnum*. The cap is 5-12 cm in diameter, soon expanded, somewhat depressed in the centre, with relatively thin and brittle flesh, of a leathery buff colour, but sometimes with a reddish flush, matt, densely felted to minutely tomentose-squamulose, with no concentric rings. The stipe is only 1-1.5 cm. thick, cylindrical, often rather long, relatively firm, hollow when older, light ochre-brownish. The gills are crowded, shortly decurrent, whitish and later ochre-yellowish. The pale ochre-pinkish or pale ochre flesh has a distinctive specific odour of soup spices and a mild taste. When cut it exudes small quantities of watery, bitterish latex not changing colour on exposure to air. The spore print is whitish.

Older mycological manuals recommend *Lactarius helvus* as spice for seasoning soups and sauces, although it is evidently poisonous. Symptoms of poisoning appear after the ingestion of several fruit bodies, usually accompanied by vomiting.

L. helvus can hardly be confused with any other species. Only *L. quietus* is slightly similar, but this smells of bedbugs, its cap cuticle is entirely glabrous, and injured flesh exudes cream-yellow, unchanging, slightly itchy latex. It is very abundant under oaks and also grows in dry places. It is harmless.

Common Earth Ball

This common inhabitant of forests, particularly sandy pine woods, may sometimes be mistaken for a truffle. However, the Common Earth-ball, together with some other closely related and similar species of *Scleroderma*, represents the class Gasteromycetes (stomach fungi). Their spores arise on basidia and not within asci, as is the case in truffles. The Common Earth-ball has tuberous or subglobular fruit bodies with a typical structure; they are formed by two layers—the outer peridium and the inner gleba. The peridium, cracking into wart-like structures, is tough and distinctly separated from the gleba. The gleba of young fruit bodies consists of a white, firm, minutely chambered context emitting an intense spicy odour. The context soon acquires a rosy hue, gradually passing over into poppyseed-blue, greyish and black. On reaching the final stage of its development, the context disintegrates into a dark green, powdery matter containing spores. These are released through openings at the apex of the irregularly cracking peridium to be dispersed in the neighbourhood.

The Common Earth-ball is a poisonous fungus. Despite this it can be used as a spice, but only in small quantities (the amount of dried powder approximately covering the tip of a rounded knife, or several slices of young fruit bodies). Overdosed foods or those prepared from ripe fruit bodies cause poisoning accompanied by vomiting and heart troubles.

Fairly common along hedgerows and roadsides is *S. verrucosum* with its thin, minutely warty peridium. It often terminates in a stipe-like and furrowed base ending in a cluster of mycelial threads.

Boletus Lupinus

This is a robust, fleshy bolete with a 6-14 cm wide cap, subglobular when young, then semiglobular to slightly convex, ash-grey to whitish in colour, sometimes with a slight greenish tint, often with pink or red patches, and tinged yellow at the margin. The tubes are yellow, then ochre-yellow, up to 15 cm long, with predominantly carmine or dingy purple-red angular pores staining blue where handled. The shortly cylindrical stipe is pale yellow above, brownish below, sometimes with a reddish flush. For the most part it bears no reticulation, but only occasionally a fine reticulum is present below the apex. The flesh is yellow but pales

in older specimens, it is brownish in the stipe base, staining blue throughout the fruit body when cut. It has an acid, rather unpleasant taste and smell, slightly reminiscent of *Lepiota crisata*.

Boletus lupinus is a very rare and hitherto little known species, related to the Devil's Boletus (*B. satanas*) which it greatly resembles. However, the stipe of the Devil's Boletus is marked with a reticulum reaching to its lower half at least, the cap is whitish or pallid brownish, never bearing red patches but often with a greenish tinge. Old fruit bodies of the Devil's Boletus emit a nauseous odour of decomposing flesh.

B. lupinus is most commonly found in the fallen foliage of beech forests in France, Italy, Switzerland, Denmark, Sweden, the Czech Republic and Slovakia. Like the Devil's Boletus, it is poisonous.

.44	2754	2761	2767
.45	2818	2825	2831
.46	2884	2891	2897
.47	2951	2958	2965
.48	3020	3027	3034
.49	3090	3097	3105

13

Characteristic Features of Fungi

Nowadays the life of fungi belongs to the most interesting chapters of mycology. Heterotropic nourishment, which represents the principal characteristic of this huge group of organisms, considered as an independent kingdom besides green plants, and animals, has called forth a number of divergent phenomena and properties.

Saprophytism and parasitism, two large categories, differing in their, relation to the environment, from which fungi obtain organic substances, necessary for building up their fruit bodies, assert themselves, in characteristic manner in the development of fungi.

In the remote past they gave rise to innumerable fungus forms, differing not only in shape but also physiologically and genetically. Their classification is a task to be coped with by mycological taxonomy.

Saprophytic fungi live in their mycelial form in an environment composed of dead plant and animal remains at different stages of decomposition.

Their hyphae penetrate through the soil humus and detritus through the wood of decaying tree trunks, stumps and branches, or live in the net work of withering leaves, needles, herb stems, grass stalks etc. Many saprophytic fungi are bound by their way of life to specific and often limited environment.

A generally widespread phenomenon is the simbiosis (co-existence) of fungi with the roots of higher vascular plants, less frequently with ferns and bryophytes; it is termed mycorrhiza. The mycelium either clothes the roots only externally (ectotrophic mycorrhiza), or penetrates inside the root tissue (endotrophic mycorrhiza). The ectotrophic mycorrhiza is easily discernible by the naked eye: the little roots are usually short but thick and often forked so that they may resemble corals in appearance. The mycelial hyphae envelop the roots like a continuous covering with protruding short surface excrescences. The endotrophic mycorrhiza, on the other hand, does not appear on the outside because the hyphae pervade the intercellular spaces of the root network. This type of mycorrhiza is more common than the former. The fungi living in mycorrhizal dependence upon the roots of live trees are not easily raised in pure cultures in a laboratory, and it is only rarely that they may be compelled to form fruit bodies. The cultivation of other saprophytic fungi—particularly lignicolous ones—presents no great problems.

The relationship between the two organisms, the green plant and the fungus, has been and continues to be subjected to a great many studies because of the considerable practical importance of the problem area concerned. Fungi of most various systematic groups live in mycorrhizal association. Mycorrhiza is particularly frequent in Basidiomycetes, but also exists in other groups (e.g. in moulds and ascomycetes). A single tree may be found to live in this kind of symbiotic relationship with several fungus species simultaneously—as a rule, however, only one of them produces fruit bodies (often only after the tree has attained a certain age). This gives rise to the heterogeneity and variegation of the mycoflora of forest growths, whose specific structure changes with the age of the

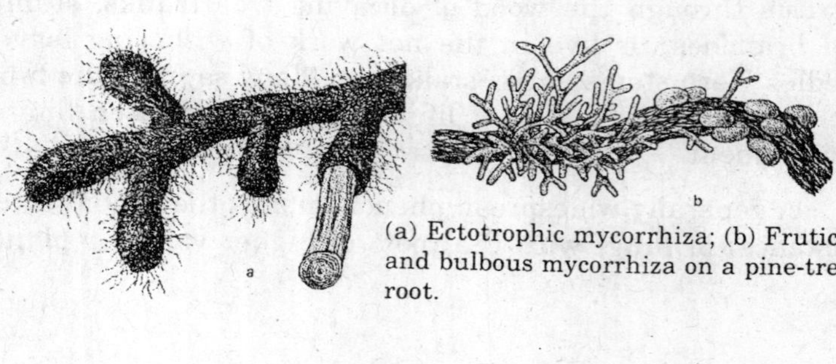

(a) Ectotrophic mycorrhiza; (b) Fruticose and bulbous mycorrhiza on a pine-tree root.

forest. Mycorrhizal fungi are indispensable for the life of some plants, e.g. for the seedlings of numerous orchids growing in free nature.

Another remarkable phenomenon is the symbiosis of fungi with animals. Termites (Isoptera) establish and grow in their nests cultures of fungi whose mycelium produces spherical bodies serving the termites as food. Various termite species grow various species of fungi as, for instance, fungi of the genus *Termitomyces* belonging to the Tricholomataceae family, whose large fruit bodies growing out of termite nests are collected by the local inhabitants for culinary purposes. Microscopic fungi are cultivated as foodstuffs by ants inhabiting trunks and stumps; the same is done by bark-beetle larvae (Ipidae) growing the mycelium of the so-called ambrosial fungi in their galleries. An independent group of the microscopic fungi Laboulbeniomycetes lives on the body surface of most various beetles, especially of those inhabiting a relatively damp environment. They do not seem to do any harm to their hosts.

In the remote past, a close co-existence of fungal hyphae and algae gave rise to complex organisms called lichens (Lichenes). Owing to the fact that the fungal component of the lichenous thallus is most frequently represented by an ascomycete, this group of lichens is called Ascolichenes (also lichenized Ascomycetes); as fruit bodies they develop apothecia or perithecia. A much smaller part of the hyphae involved here belongs to the Basidiomycetes—this group of lichens is called Basidiolichenes (or lichenized Basidiomycetes). Some mycologists now regard various gill-fungi species as lichenized Basidiomycetes, although their fruit bodies differ from those of other gill fungi is nothing but the presence of algae in the tissue of some part of the thallus.

After exhausting the nutritional substances in one place, the mycelium of some fungi starts spreading in stripes or in more or less regular circles into the neighborhood, and its growth in a particular direction manifests itself also by the development of fruit bodies. This is most conspicuous in the so-called 'fairy rings' formed by fungi of the genera *Ramaria,*

Hydnum, Clitocybe, Tricholoma, Marasmius. The shape of the rings depends not only on the nutrient content of the soil but also on the layout of tree roots, and it may also be affected by the pattern of water dropping from tree crowns. The radial growth of the mycelium can also be seen in the fungi raised in pure cultures on agar soils.

To a considerable extent, saprophytic fungi participate in the decomposition of cellulose in the wood pervaded by the mycelium which not infrequently causes its characteristic disintegration (for example, cubiform red rot, fibrous white rot, etc.). Some saprophytic wood-destroying fungi also pass over to weakened but still living trees, speeding up their death; consequently they appear as occasional facultative parasites and are called saproparasites. They enter the trunks either through frost cracks or through scars left by broken-off branches, eventually by injuries connected with wood production. The wood infested with fungi sometimes acquires a specific colour. For example, the beautifully green-coloured wood of beeches or other deciduous trees is pervaded by the mycelium of the minute disc fungus *Chlorosplenium aeruginosum* secreting xylochloric acid insoluble in water. Another well-known feature is the phosphorescence of wood created—besides by bacteria—by the mycelium of the Honey Fungus (*Armillaria mellea*). Fungi whose fruit bodies give off a bluish sheen visible by night are widespread in the tropics. The species *Omphalotus olearius* with fluorescent gills, growing in southern Europe, belongs to this group.

Other saprophytic fungi have become specialized on ecologically narrowly confined environments and are never found elsewhere. Anthracophilous fungi live exclusively on burned-over soils or in old fire-places, in fact wherever wood was burned and where, for some time at least, a suitable microclimate (soil humidity, shade) exists. In such an environment, the gradual succession of individual species can best be examined. As a rule, one of the pioneer species is the disc fungus *Pyronema confluens,* which is minute but still conspicuous due to its proliferation; it may often be seen covering the burned soil with extensive, pink-red layers of its innumerable apothecia. Somewhat later it is replaced by other representatives of disc fungi (*Geopyxis carbonaria)* and various cup fungi (*Peziza subviolacea, P. echinospora*). The last ones to grow here, when the burned-over areas are already overgrowing with mosses or liverworts, are the gill fungi *Pholiota carbonaria, Tephrocybe carbonaria,* and others.

A characteristic fungus community grows on excrements of various animals. These fungi, termed coprophilous, occur in great abundance and probably come from all fungus groups. The most conspicuous among them are members of the genus *Coprinus* whose short-lived, quickly developing fruit bodies are very common indeed. Coprophilous fungi can be successfully cultivated and studied in laboratory conditions on excrements collected in nature and kept in a moist but airy environment. They are therefore often raised as models for physiological and genetic experiments.

Some fungi are parasites, obtaining their nourishment from the bodies of other living organisms, while the host upon which they live is at a disadvantage, being more or less damaged or gradually completely destroyed by the fungus. Most of them are microscopic fungi surpassing the saprophytes in the number of species. Rusts (Uredinales) and smuts (Ustilaginales) are large groups of exclusively parasitic fungi; their economic relevance is due to the serious damage many of them cause in the tissues of cultivated plants. Rusts either develop on a single host, or change their hosts in which they successively produce various stages. Other fungi live inside their host's tissue as endoparasites, while ectoparasites live on its surface, sending within only their mycelium. There are many parasites among the Ascomycetes (e.g. mildew—Erysiphales) and Pyrenomycetes. The proportion of parasites among higher fungi is substantially lower, but cases of a large fungus parasitizing another large fungus are known. By way of example let us mention *Xerocomus parasiticus* living on *Scleroderma citrinum,* or *Volvariella surrecta* parasitizing the fruit bodies of *Lepista nebularis.* Species belonging to the genus *Nyctalis* are commonly found growing on *Russula; Cordyceps ophioglossoides* lives on the subterranean fruit bodies of *Elaphomyces*. Other numerous representatives of the genus *Cordyceps* are widespread, especially in the tropics, and parasitize insects.

There also exist huge numbers of microscopic fungi parasitizing other fungi. In autumn the fruit bodies of some boleti are covered with conspicuous golden-yellow, farinaceous layers of the hyphomycete *Sepedonium chrysospermum:* this is the conidial stage of *Apiocrea chrysosperma*. Fine moulds of the genus *Spinellus* often live as parasites on the caps of fungi of the genus *Mycena,* and brown, matted-tomentose covers of the mould *Sporodinia* occur on the caps of various other gill fungi. However, parasites also have their parasites—for example, the rust *Puccinia* is attacked by the coelomycete *Darluca filum*.

An ecologically specific group are fungi confined to an aquatic environment. Most abundant are the representatives of water moulds (Blastocladiales, Saprolegniales, etc.) living on decaying organic matter submerged in water. Some of them even parasitize aquatic animals. Still smaller aquatic moulds live as parasites on pollen grains fallen into the water, or within the cells of water algae. Rotten leaves lying in shallow and clear water currents are inhabited by a whole community of microscopic Hyphomycetes producing conidia of extraordinary shapes. Larger fungi growing on decaying wood logs and branches submerged in water include representatives of Pyrenomycetes as well as some Discomycetes. A quite specialized group consists of some Pyrenomycetes growing on wood submerged in sea water.

Both parasitic and saprophytic fungi trigger off various chemical changes in the substrate or in their host's body through the operation of their enzymes (ferments). The most important of these, which are most utilized by man, are enzymes evoking alcoholic fermentation. Also amylolytic (disstatic) ferments transforming starch into sugar are generally applied.

Developmentally a very old group of organisms, fungi are at home almost all over the Earth, and their spores are practically omnipresent in the atmosphere. This of course does not imply that the distribution of all the species is worldwide (cosmopolitan). Many species of fungi have an extensive range of distribution, but far more species are confined to specific, often considerably restricted, localities. The occurrence of such fungi is conditioned not only by the presence of an adequate substrate or host, but also by climatic factors, by the underlying geological stratum and its chemical composition, etc. Furthermore, the structure of the mycoflora of each region is the result of a long-lasting developmental process. While the genera and species of fungi growing in many European countries are almost the same, in North America, where the number of species is considerably higher, only about one-third of the total corresponds to European species. One reason for this is the specific abundance of woody plants and herbs in North America which have affected the structure of fungus species in the course of their long-term development. While the temperate belts of both hemispheres are characterized by a predominance of saprophytic fungi, in the tropical regions there is an exceeding abundance of microscopic parasitic fungi. It should be emphasized that not all the parasites closely dependent on their host are present in every locality inhabited by that host: their occurrence is often substantially restricted, or they are known from isolated finds only; they may also pass over to substitute hosts. As regards the elevation above the sea level, in comparison with other vascular plants no striking differences appear in the fungi; yet a great many species occur exclusively in high mountains (even above the timberline), others again grow at lower elevation only.

Adventive fungus species are brought from their original homes to remote regions, e.g. by the action of man. A classical example is the transfer of some stinkhorns (Phallaceae) from North America and Australia to Europe.

